Recent Advances in Nanoscience and Technology

Editors

Sunil Kumar Bajpai
Murali Mohan Yallapu

eBooks End User License Agreement

CONTENTS

FOREWORD

Metal nanoparticles play an interesting role both in nature as well as in the area of nanotechnology. With the number of publications mushrooming in the different areas of nanotechnology not only in the core areas of nano-materials, but meanwhile also in many specialized areas, this monograph on nanoparticles is filling the existing gap in the area of nanoparticles, above all in metal nanoparticles.

The rapid progress in the area of nanoparticles in general and in metal nanoparticles specifically absolutely justifies a monograph in this expanding field. The various applications of nanoparticles make them very useful materials or precursors for many processes. Interestingly, the great variety of nanoparticles is not yet reflected in the literature accordingly in this currently very active research field on a global scale. However, one is able to sense the trends and anticipate future directions in this exciting area.

This book covers a variety of different topics in this novel area that are reflected in the chapters and is focusing predominantly on metal nanoparticles. The first chapter is devoted to the synthesis of nanostructures in aqueous media. In the second chapter, surfactant-based synthesis methods for metal nanosystems are covered. In the third chapter the microemulsion-mediated synthesis of nanoparticles is considered. Chapter four handles the synthesis of metal nanostructures by photoreduction. The general aspects of self-assembly of nanostructures are discussed in chapter five. The core topic of the monograph, metal nanoparticles, is contained in the sixth chapter. There, the different synthesis approaches to metal nanoparticles and polymer metal nanocomposites are highlighted. Finally, in the last chapter, three-dimensional networks based on hydrogels are presented as nanocarriers for metal nano-particles.

In conclusion, this monograph is a valuable source for scientists, researchers, and students, who are interested in the area of metal nanoparticles and the latest developments in this interesting and important field.

Kurt E. Geckeler, M.D., Ph.D.
Chair, Department of Nanosystems Engineering (DNE),
World-Class University Program (WCU)
Professor of Materials Science and Engineering
Professor of Medical System Engineering
Gwangju Institute of Science & Technology
KOREA

PREFACE

The word "nano" has become very popular throughout the world and it has brought revolution in the traditional concept on material dimensions. Nanomaterials are defined as materials that have dimensional size on a nanometer scale of 1-100 nm and exhibit distinct physico-chemical properties than bulk materials. Scientists, to some extent, believe that nanoparticles constitute another state of matter! Nanostructure materials resolve a number of common problems associated in the fields of industrial, engineering and biomedical sciences. These materials are being utilized in various applications including bio-markers, solar cells, electronic devices, advanced ceramics, new batteries, engineered catalysts, functional paint and ink, magnetic resonance imaging, targeted drug delivery, and lighting technologies, etc. As nanoscale devices have become more of a commercial reality, the industrialization of nanoscale materials has been limited by the need for new material compositions and the development of high-throughput automation for materials preparation.

The formation of nanocrystals is notorious for its complexity and requires long reaction times, in the order of hours. However, such large-scale reactions exhibit inhomogeneities in the growth process by thermal process, which produce poor nucleation and therefore broaden size distributions. Chemical solution methods have been widely used to produce nanostructured materials, and can be applied to achieve monodisperse nanoparticles with controlled size and shape. There is no general strategy to make nanoparticles with narrow-size distribution, tailored properties, and desired morphologies, which could be universally applied to different materials. It is believed that nanoparticle formation follows the classic LaMer mechanism, which suggests a short burst of nucleation followed by slow diffusive growth, favoring formation of nucleation followed by slow diffusive growth, thereby favoring formation of monodisperse crystalline nanoparticles. However, the regulation of the size feature, surfaces and interfaces are crucial components in the synthesis of nanostructured systems. Each specific synthetic route of nanoparticles dictates their usage for a particular filed of application. Therefore, identifying novel methodologies to prepare monodisperse nanoparticles by utilizing natural resources and those developed products can directly be applicable for bio-medical applications due to no harmful reagents involved in the preparation processes.

The present book deals with various advanced strategies, such as, surfactant based synthesis, microemulsion mediated synthesis, self-assembly process, polymer and hydrogel template synthesis, and natural resources based synthesis, that have frequently been followed to fabricate nanostructures of required size and shape, and functionalities to enable them to be used in a wide spectrum of industrial, biomedical and technological applications. It is intended to give readers a clear picture of nanoparticles synthesis by various methodologies as well as new ideas or suggestions on the creation of novel nanostructure materials to improve the performance of the advanced functional nanomaterials.

S.K. Bajpai
Govt. Model Science College, Jabalpur, India
Murali Mohan Yallapu
Sanford Research/USD, Sioux Falls, USA

CONTRIBUTORS

Alexander Pyatenko	Scientist, National Institute of Advanced Industrial Science and Technology (AIST), Japan, 2-17-2-18, Tsukisamu-Higashi, Toyohira-ku, Sapporo, Japan
Bo Hu	Postdoctoral Fellow, Division of Nanomaterials & Chemistry, Hefei National Laboratory for Physical Sciences at Microscale, the School of Chemistry & Materials, University of Science and Technology of China, P. R. China.
Deepa Sarkar	Postdoctoral Fellow, Department of Chemical Engineering, Indian Institute of Technology, Bombay, Powai, Mumbai
Hanying Zhao	Full Professor, Key Laboratory of Functional Polymer Materials, Ministry of Education, Department of Chemistry, Nankai University, Tianjin, 300071, P. R. China
Hong-Yan Shi	Graduate Student, Division of Nanomaterials & Chemistry, Hefei National Laboratory for Physical Sciences at Microscale, the School of Chemistry & Materials, University of Science and Technology of China, P. R. China.
Jian Zhang	Graduate Student, Key Laboratory of Functional Polymer Materials, Ministry of Education, Department of Chemistry, Nankai University, Tianjin, 300071, P. R. China
Kartic. C. Khilar	Full Professor, Department of Chemical Engineering, Indian Institute of Technology, Bombay, Powai, Mumbai
Manjula Bajpai	Full Professor, Department of Chemistry, Polymer Research Laboratory, Govt. Model Science College, Jabalpur, MP 482001, India
Mohana R. Konduru	Full Professor, Department of Polymer Science & Technology, Sri Krishnadevaraya University, Anantapur, Andhra Pradesh, India
Mustafa Çulha	Associate Professor, Yeditepe University, Faculty of Engineering and Architecture Genetics and Bioengineering Department, Kayisdagi, Istanbul, Turkey
Murali M. Yallapu	Research Fellow, Cancer Biology Research Center, Sanford Research/USD, Sioux Falls, USA
Samba S. Kuruva	Postdoctoral Researcher, Department of Polymer Science & Technology, Sri Krishnadevaraya University, Anantapur, Andhra Pradesh, India
Shu-Hong Yu	Full Professor, Division of Nanomaterials & Chemistry, Hefei National Laboratory for Physical Sciences at Microscale, the School of Chemistry & Materials, University of Science and Technology of China, P. R. China.
Sunil Kumar Bajpai	Full Professor, Department of Chemistry, Polymer Research Laboratory, Govt. Model Science College, Jabalpur, MP 482001, India
Varsha Thomas	Postdoctoral Fellow, Department of Chemistry, Polymer Research Laboratory, Govt. Model Science College, Jabalpur, MP 482001, India

CHAPTER 1

NANOSTRUCTURE SYNTHESIS IN AQUEOUS MEDIA

Alexander Pyatenko

National Institute of Advanced Industrial Science and Technology (AIST), Japan, 2-17-2-18, Tsukisamu-Higashi, Toyohira-ku, Sapporo 062-8517; Tel: +81-44-857-8443; Fax: +81-11-857-8984; E-mail: alexander.pyatenko@aist.go.jp

Abstract: In this chapter the main experimental methods used for the synthesis of metal nanostructures in aqueous colloidal form are considered. For convenience, consideration was made separately for noble metals and magnetic metals, even the most of methods are generally common for both groups. While the chemical reduction of metal ions still remains the main method for such synthesis, laser ablation in water solutions as well as combinations of these two methods became more and more popular, and can bring more promising results in the nearest future.

Key words: Stabilization, chemical reduction, colloid nanoparticles, surfactants.

1. INTRODUCTION

The interest to nanoscience and nanotechnologies grew dramatically and the amount of different objects of such nano researches and applications increased exponentially in the last decade. The former titles of books were "Introduction in Nanotechnology" or "Nanoparticles", but nowadays, it is impossible to observe even in a single large book such wide aspects. The subject of this book is metal nanostructures synthesis and the particular subject of this chapter is the synthesis of metal nanostructures in aqueous (or water) solutions, or colloidal nanostructures' synthesis. Even for such particular purpose we need to define first what kind of metal nanostructures and synthetic methods will be considered here. First, we will consider here only the primary synthetic processes. Usually, the result of primary synthesis is known as nanoparticles. Following this tradition can be considered as the product of synthetic process "nanoparticles", taking into account the sizes and shapes of different nanoparticles which can be extremely different (their sizes can be varied from few nm to μm, and their shape can be varied from spheres to nanowires). These nanoparticles then can be used as building blocks for larger supramolecular units, but such constructions are not the subject of our consideration. The wide class of nanoparticles, semiconductor nanoparticles or quantum dots will also be considered in other chapters. All of metal nanostructures or nanoparticles considering in this chapter can be divided into two main groups: noble metals nanoparticles, and magnetic nanoparticles. In spite of the fact that generally the same methods are used for synthesis of different types of nanoparticles, many characteristic features exist when metal particles of different group are synthesized. For convenience, we have discussed the synthesis of particles of two main groups separately.

We discuss here the particle synthesis in aqueous media only. The synthesis in non aqueous solutions as well as the synthesis with the surfactants or polymers will be discussed in other chapters. We observed that colloid consisted of the naked nanoparticles in pure water is unstable normally due to the particles agglomeration. To prevent the agglomeration process, colloidal particles need to be stabilized. There are two methods of particle stabilization, electrostatic (Coulombic repulsion) and steric (polymeric or other organic "overcoat") [1-3]. In the first method, the particles were prevented from agglomeration by electrical double layer formed by negative ions absorbed on the particle surface, and positive charges induced on the metal particle surface by mirror effect (see Fig. **1**).

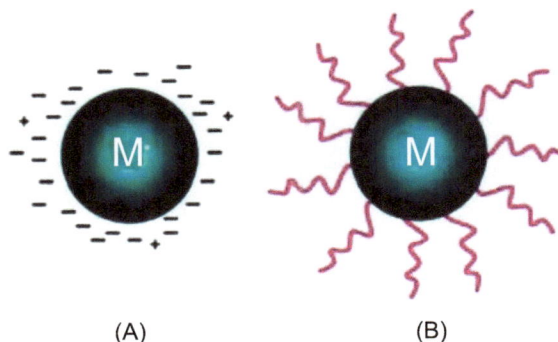

(A)   (B)

Fig. (1). Schematic representation for (A) an electrochemically stabilized metal (M) particle (i.e., one stabilized by the absorption of ions and the resultant electrical double layer) and (B) a satirically stabilized metal particle (i.e., one stabilized by the absorption of polymer chains).

In the second method, the particle agglomeration was prevented by absorption of polymer, surfactant or ligand molecules at the surface of the particle. Recently, a new approach for particle stabilization was developed. In this approach, particle surface is covered by a thin layer of SiO_2, thus metal core- silica shell particles were formed [4-5]. Here we observed that if surfactants, polymers or other molecules not soluble in water were used just for the particle stabilization and they do not play essential role in synthetic process itself.

2. NOBLE NANOPARTICLE SYNTHESIS

Nanoparticles (NPs) of noble metals and particularly of gold were the first nanoobjects of synthesis and study [6-7]. Long time back, there were only spherical or quasi-spherical particles. More than ten years ago, nanorods [8-13], and nanowires [13] became the second and the third nanoobjects. Recently, different types of nanoparticles have grown dramatically. Until today nanocubes [14-16] and nanoprisms [16-18], nanoplates [19-21], and nanobelts [22, 23], flower-shape [24-26], and many other shapes of NPs have been synthesized. Some of these nanostructures are essentially two-dimensional, like nanodisks, or nanoplates, and some are even one-dimensional, like nanowires. Earlier, only gold nanoparticles [7] were produced in colloidal form. After that, silver [27], copper [28 -33], and platinum [34-38] NPs were added to the noble NPs family. Nowadays, all other noble metal structures, such as palladium, rhodium, rhenium, osmium and ruthenium NPs can also be synthesized in colloidal form successfully [31, 39-44].

We now consider the experimental methods which are being used for these particles synthesis in aqueous solution.

2.1. Chemical Reduction

Chemical reduction of noble metal ions dissolved in water by adding the reducing agent in aqueous solution is the oldest and still the most common method for preparing the colloidal particles. For the reduction of metal ion by reaction of gold it is normally chloroauric acid, HAuCl4, or sodium or potassium salts of this acid used as the source of $[AuCl_4]^-$ ions (complex of Au^{3+} with four Cl^- ions).

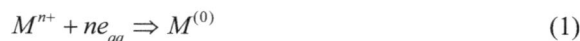

$$M^{n+} + ne_{aq} \Rightarrow M^{(0)} \qquad (1)$$

The reducing agent, X^m, must be oxidized by the reaction

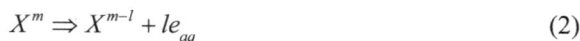

$$X^m \Rightarrow X^{m-l} + le_{aq} \qquad (2)$$

In case of silver, silver nitrate, $AgNO_3$, is the most common source of silver ions, Ag^+, although silver sulphate, Ag_2SO_4 [45], or silver perchlorate, $AgClO_4$ [46] have also been used. Chloroplatinic acid, HPtCl4, hexachloroplatinate, H_2PtCl_6, as well as salts of these

acids are normally used in case of platinum. For other noble metals, chlorides, like $CuCl_2$ or $PdCl_2$, sulfates, nitrates, or some other salts of these metals can be applied.

Sodium citrate [7, 45-48], sodium borohydride, NaBH4 [45, 47-52], ethanol [53-55], and propanol [56], ethylene glycol (EG) and poly(ethylene) glycol (PEG) [57-59], ascorbic acid [22, 24, 60-65], poly(*N*-vinylpyrrolidone), (PVP) [66], glycine [67], aniline [68], hydrazine [61, 69-71], mono- and disaccharides [16, 72] have all been used as reduction agents, for e.g. Citrate [73-76], PVP [24, 53, 62, 66, 79], PEG [57], thiol terminated molecules [50, 80-85] as well as different surfactants like NaAOT [49, 86-87], CTAB [21, 64, 88-92], CTPAB [93], or STEAB [93] which are used as stabilizers or protecting agents against particles aggregation. To improve the synthetic process especially to make the size and shape control of the particles produced more precise, many researcher are still searching and finding new reduction agents and stabilizers.

Reduction of chloroauric acid, HAuCl4 by sodium citrate was the first experimental method developed in detail by Turkevich with coworkers in 1951 [7] for the synthesis of colloidal gold nanoparticles. This procedure is used very commonly as a standard or classical citric reduction method. Applying for silver, this method was developed by Lee and Meisel in 1982 [45]. This method consists of the adding of appropriate amount of sodium citrate to boiling water solution of HAuCl4 (in case of gold) or $AgNO_3$ (in case of silver). The reaction than continues with extensive stirring during one hour. No stabilizing agent is needed. Turkevich [7] also studied the mechanism of the reduction process, and showed that the reduction of ions accomplished the oxidation of citrate to acetone dicarboxylate (ADC) as it is shown in Scheme 1.

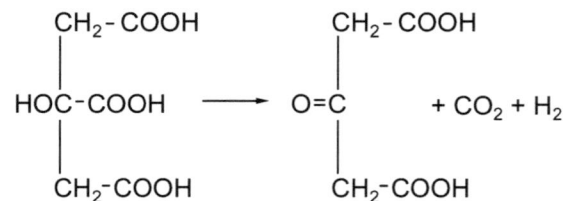

$$
\begin{array}{ccc}
CH_2-COOH & & CH_2-COOH \\
| & & | \\
HOC-COOH & \longrightarrow & O{=}C \qquad + CO_2 + H_2 \\
| & & | \\
CH_2-COOH & & CH_2-COOH
\end{array}
$$

or in ionic form:

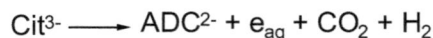

$$Cit^{3-} \longrightarrow ADC^{2-} + e_{aq} + CO_2 + H_2$$

Scheme 1

Using citrate reduction method, it is rather easy to produce relatively large (tenth to hundreds nanometers) particles in relatively large concentrations. In case of gold, the particles produced are spherical or quasi-spherical with rather narrow size distribution of about 10-20 nm [94].

But in case of silver, the situation is not so good. Silver nanoparticles produced by citrate reduction method are essentially larger, varied rather widely in their sizes and shapes [95] (see Fig. 2), and additional efforts are required to control their shape or to produced monodisperse spherical particles.

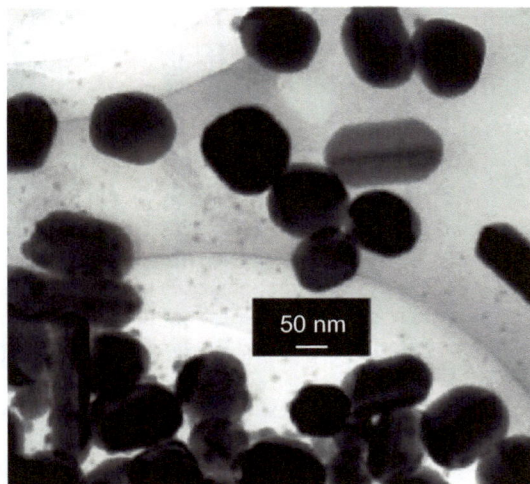

Fig. (2). TEM image of silver nanoparticles prepared by classical citric reduction method. (Average size of the particle is about 100 nm). Images reprinted with permission from Ref. [95] Copyright © 2005 American Chemical Society.

Another very important reduction agent is BH_4^-. Sodium borohydride reduction method was developed by Creighton with coworkers [47] Lee and Meisel, [45] and now used widely as a citric reduction method. Compare to citric reduction where water should be at boiling temperature, borohydride reaction applies ice cold water conditions. In a citric reduction, additional stabilizing agent is needed. Once the reduction was completed, the solution was boiled for one hour to decompose any excess of NaBH₄. The great advantage of this method is the synthesis of small (few nanometers), rather monodisperse spherical particles.

Two different processes, particle nucleation (bulk reduction of M^{n+} ions) and particle growth (reduction of M^{n+} ions on the surface of already formed particles) coexisted in chemical reduction process. As a result of this coexistence, particles of different sizes and shapes are synthesized. For more precise control of the synthetic process, it would be better to separate these two processes temporarily. This is the idea of seed method [12, 96-99]. First, the seed particles are prepared by using a bulk chemical reduction or other technique, after that these seed particles are used as the nuclear for further particle growth. When the seed method is used for the synthesis of larger spherical nanoparticles from smaller spherical seeds, then this assumption leads to a very simple equation, connecting the diameter of seed particles, d_{p0} with the diameter of finally synthesized particle, d_p

$$\frac{d_p}{d_{p0}} = \left(1 + \frac{n^+}{n_s}\right)^{1/3} \tag{3}$$

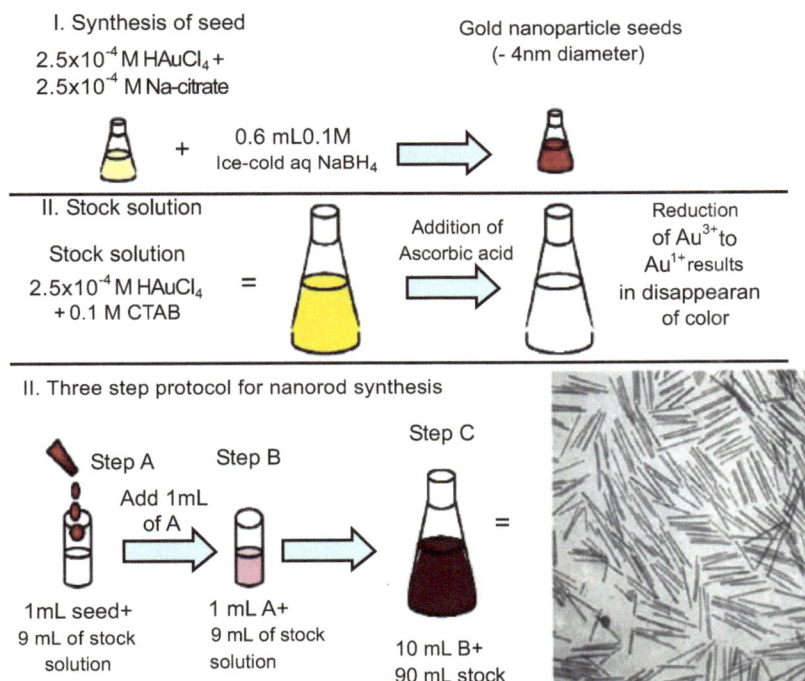

Fig. (3). Schematic preparative route of CTAB stabilized silver and gold nanoparticles. Images reprinted with permission from Ref. [103] Copyright © 2005 American Chemical Society.

where n_s and n^+ are the mass concentrations of seed particles and M^{n+} ions added to the seed particle colloid [43, 100]. Jana with coworkers successfully produced spherical monodisperse gold nanoparticles of different sizes using this method with citrate reduction [98-99]. The main assumption of this method, that the reduction of M^{n+} ions occurs only on the surface of seed particles, is valid only to some extent. The bulk nucleation can be small but not negligible. To neglect the bulk nucleation, the relative amount of M^{n+} ions in a colloid must be small comparing with the amount of seed particles, which means that the ratio n^+/n_s must be small. But, according to equation (3), if this ratio is very small, the effect of particle growth will be negligible. Thus, to increase the particle size essentially, the relatively large amount of M^{n+} ions (with reduction agent) should be added to seed colloid, and this is how the bulk nucleation will take place, increasing the particle size distribution. As a compromise to these controversial demands, the particle growth is accomplished in this method in several consecutive steps [98-99].

Recently, seed method has been successfully used to synthesize noble nanoparticles of different shapes. The main idea is to suppress the particle growth in some directions, and therefore, to make the particle growth inhomogeneous. For this purpose specific interaction of stabilizing agent with different growing faces of seed particles is applied. For the synthesis of gold and silver nanorods this method was developed in detail by Murphy and coworkers [101-102] and later on summarized in a feature article [103]. Reduction of $HAuCl_4$ or $AgNO_3$ accomplishes on the surfaces of small silver or gold seed nanoparticles, formed preliminary by citric or borohydride reduction, in the presence of ionic surfactant, cetyltrimethyl-ammonium bromide (CTAB). Details of seed mechanism approach developed by this group are shown in Fig. (3).

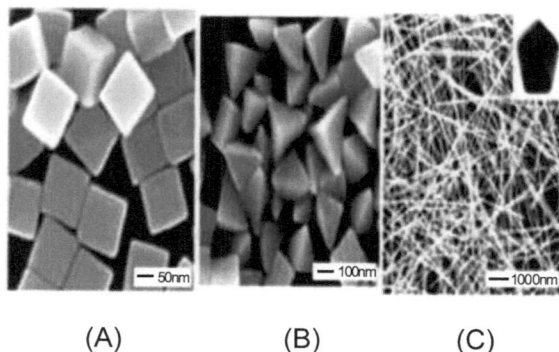

(A)   (B)   (C)

Fig. (4). SEM image of (A) nanocubes, (B) nanopiramides, and (C) nanorods. Images reprinted with permission from Ref. [17] Copyright © 2006 American Chemical Society.

Other approach is used by Xie and coworkers for shape- controlled synthesis of silver nanoparticles [14,

17, 78, 104-105]. First using the improved polyol reduction method [106], where liquid polyols such as ethylene glycol both used as the solvent medium and reduction agent, they synthesize either twinned- or single- crystalline seeds. Next, using PVP reduction of additional silver nitride or chloroaurate on the different faces of seed particles, they successfully produce nanoparticles of different shapes. NPs with different shapes (cubes, prisms, wires) synthesized by this method are depicted in Fig. (4).

Among different chemical reduction techniques, one worthy technique is silver oxide reduction by hydrogen gas in water, although generally considered insoluble in water; Ag_2O has a low solubility of 0.053 g/L at 80 °C, which is sufficient for nanoparticle synthesis. Silver nanoparticles are synthesized in a saturated silver oxide solution at elevated temperatures in equilibrium with hydrogen gas at elevated pressure.

Because hydrogen, water, and silver oxide are the only components used in the reaction, no other chemical was presented in the final colloidal suspension. One of remarkable results of this synthesis is that the particles contain no foreign stabilizers or any ions other than those from silver and water. In spite of that colloids were stable and no changes in particle sizes were observed for months and even one year. The authors explained this extraordinary result by strong coordination of water to the silver surface. Monodisperse particles of different sizes, from 30 nm to 140 nm were produced by this method. The results of this synthesis very strongly depend on any impurity existed in the water or Ag_2O, as well as on interior glass surface of the reaction vessel. For example, the results obtained with the quarts vessel were completely different from the results obtained with Pyrex vessel.

Among the different particle stabilization techniques, application of thiol terminated molecules looked very promising for further applications of those nanoparticles in bio science and bio technology. These molecules can play double role, to prevent particle from agglomeration and to be linkers between particle and bio object, such as protein, peptide, DNA fragment or even whole cell.

2.2. Laser Ablation in Aqueous Solution

There are different experimental techniques where colloidal nanoparticles produced by interaction with different forms of electro-magnetic radiation, like a visible light, UV-, or γ- radiation [108-112]. In photolysis or radiolysis technique the reduction of Me^+ ions occurs by solvated electrons or free radicals [108, 113-115]. These methods will be discussed in another Chapter. Here we consider another method of nanoparticle synthesis in aqueous solution, which used the metal surface interaction with particular form of electromagnetic radiation, short pulse of laser irradiation.

Laser ablation in gas phase is a very well known method which has been used successfully for many years [116-119]. The idea of applying this technique for liquid phase was proposed by Cotton-Chumanov group [120-121] and Henhlein group [122] in 1993. Further, this method was developed in detail by Mafune and coworkers [123-130]. Typical experimental scheme is depicted in Fig. (**5**).

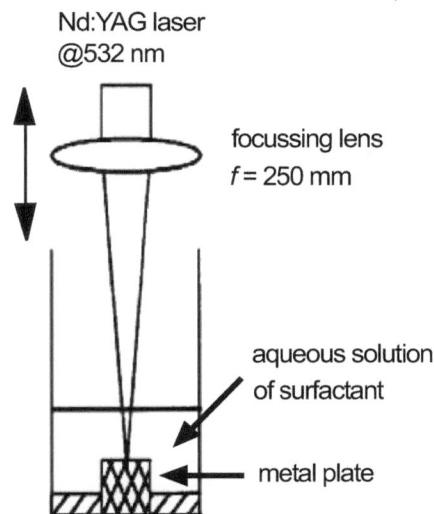

Nd:YAG laser
@532 nm

focussing lens
f = 250 mm

aqueous solution
of surfactant

metal plate

Fig. (5). Schematic diagram of the experimental apparatus. Images reprinted with permission from Ref. [124] Copyright © 2000 American Chemical Society.

Surface of the metal plate immobilized in water solution is irradiated by laser beam with different parameters (wavelength, pulse duration and pulse energy, pulse repetition rate). Lens is used usually for the beam focusing on the metal surface. To make the process more homogeneous, some researchers rotate the metal plate or vessel during an ablation [131-133]. The following mechanism of nanoparticle formation was proposed by Mafune *et al.* Absorption of laser pulse energy causes the plume formation, small cloud of hot dense plasma over metal surface contained high concentration of metal atoms and ions. Metal atoms in the plume aggregate rapidly into small embryonic particles (or nuclears) as fast as metal atoms collide mutually. After that, two concurrent processes take place: particle growth and particle stabilization. Final average size of the particle depends on chemical composition and concentration of stabilizer. By using this method, it is rather easy to produce relatively large concentration of nanoparticles with different sizes. Similar to chemical reduction technique, it is necessary to chose very carefully the chemical composition and concentration of stabilizer to produce the particles of desirable size. Average size of the particles and particle size distribution depend strongly on the comical composition and concentration of the

surfactant [124-125, 127, 129-130, 134-136]. The results of ablation experiments also depend on laser wavelength [137-138], laser fluorescence [124,129,131,135,139-141], beam focusing [120, 131, 142], duration of ablation process [124, 129, 131, 142-143], and mixing conditions [137]. This method was used successfully for the synthesis of nanoparticles, practically for all noble metals such as, gold [120, 123, 127-128, 134-136, 138-141, 143], silver [120, 124-125, 131-133, 137-143], platinum [120, 129-130], copper [120, 141], palladium [120, 144]. Compare to chemical reduction technique, this method is relatively expensive because of the laser cost, especially if a femtosecond laser is used. Pyatenko and coworkers [95] used this method for the synthesis of small silver nanoparticles in pure water without any stabilizing agents. They applied Nd: YAG laser with high power (0.34 J/pulse) and high laser focusing conditions (spot size of laser beam on the silver surface was about 0.6 mm, when initially diameter of laser beam was 7 mm). Colloids were stable in the absence of any stabilizer like the colloids obtained by chemical reduction of silver oxide with hydrogen [107] (see above), that could be explain by strong coordination of water to the silver surface. Similarity between these two cases is low concentration of silver particles into the colloids. Maybe, this low concentration makes possible to stabilize the particles by hydroxide ions only.

Since the colloidal particles produced as a result of metal ablation target can be irradiated by subsequent laser pulses. The secondary process, the process of particle fragmentation and size reduction will occur simultaneously with primary ablation process [135, 145]. It makes the process more complicated.

Recently, Pyatenko with coworkers [145] proposed a new method, which combined the chemical reduction with laser treatment technique. First they synthesized the seed particles irradiating the silver colloid prepared by citric reduction method by high intensive laser beam. The seed particles were spherical and monodisperse with average size of 8 +/- 1.7 nm [131]. Next they try to use these seed particles for subsequent growth using the usual procedure of seed method. But the growth of spherical particles accompanied by the synthesis of nanorods. Finally, to synthesize only spherical particles of different desirable sizes, the authors [145] proposed to use multi-step procedure with laser treatment of intermediate colloids. As a result, the spherical monodisperse silver nanoparticles with different average diameters from 10 to 100 nm were synthesized [145]. Some results of these synthesizes are shown in Fig. (**6**).

This method can easily be transfer to other metals, but cannot be applied for production of particles of different (not spherical) shapes.

2.3. Other Methods

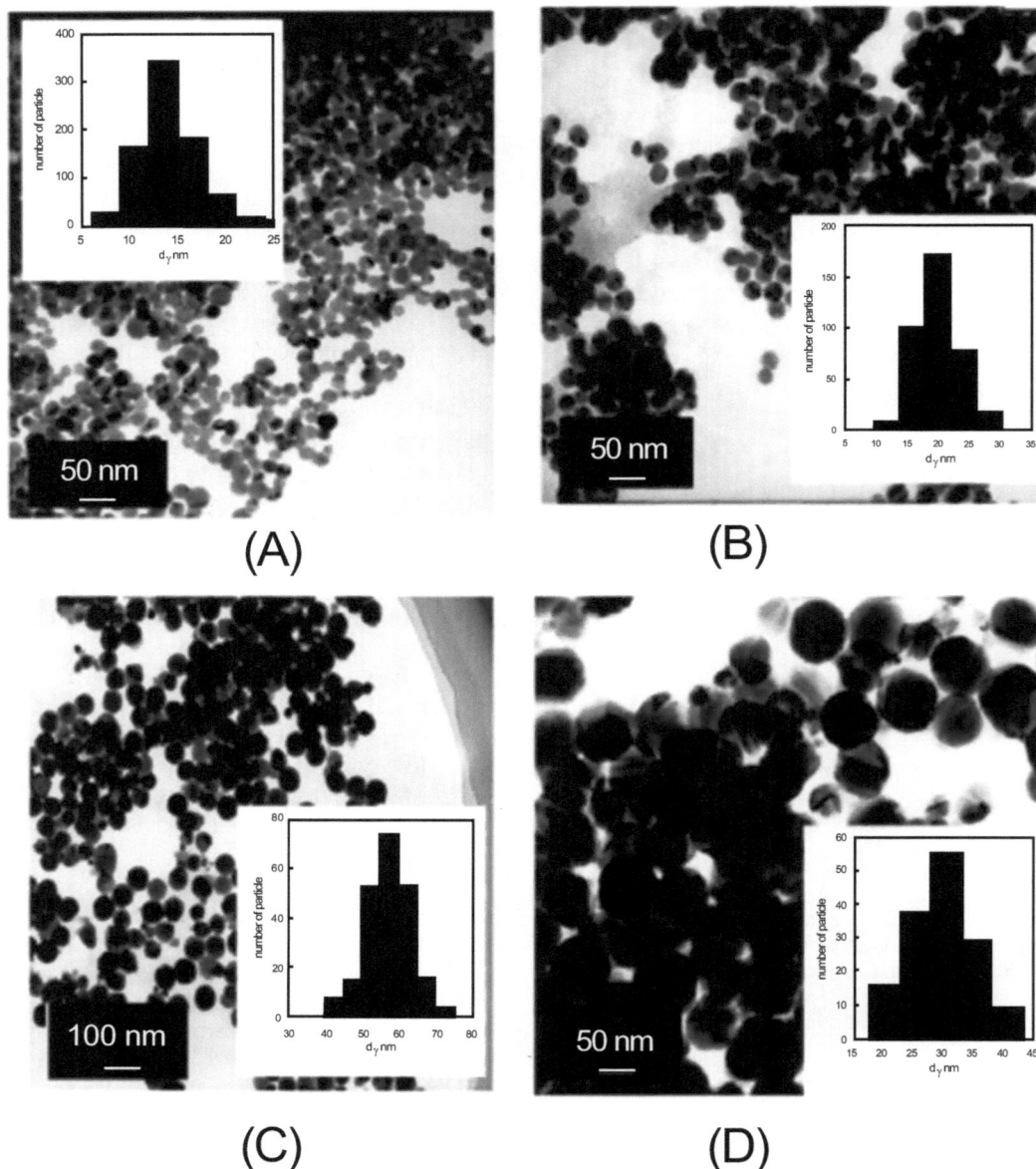

(A)   (B)   (C)   (D)

Fig. (6). TEM image of spherical nanoparticles with average diameter 14.2, 20.2, 57.2, and 82.7 nm. Insets are the histogram of particle size distribution. Images reprinted with permission from Ref. [145] Copyright © 2007 American Chemical Society.

In a sonochemical technique the energy of cavitation is used for nanoparticle synthesis. The sound of high frequency (from 20 kHz to several MHz) is transmitted through a medium as a pressure wave and induces a vibrational motion of the molecules through it. This ultrasonic treatment initiates acoustic cavitation, i.e., the formation, growth, and collapse of microbubbles within a liquid. The collapse of these bubbles locally heats up the media to thousands of Kelvin [146-147]. These extremely high temperatures make ultrasound a powerful tool for thermal chemical reactions inside the bulk solution under room temperature. First the stable colloid silver nanoparticles in aqueous solution of silver nitrate under ultrasonic treatment were produced by Nagata *et al.* [148] in 1992. Colloidal gold particles in aqueous

HAuCl4 were prepared by Grieser with coworkers in 1993 [149]. Nowadays, nanoparticles of different metals with different size, shape, and structure were synthesized using this technique [150-153]. A number of factors influence cavitation efficiency and the chemical and physical properties of the products. The dissolved gas, ultrasonic power and frequency, temperature of the bulk and type of the solvent are the important factors that control the yield and properties of the synthesized nanoparticles. For the more precise control of particle size and shape, this technique can be successfully combined with seed technique [154].

Recently, many researchers used micro wave technique instead of traditional heating of aqueous solution of reagents. NPs of different noble metal, Au,

Ag, Pt, Pd [16,155-157] with different sizes [71, 158] and shapes [16, 159] were synthesized already using this approach. Many authors point out the fast heating as the main advantage of this technique. Different researchers use different MW power (from 300W [158] to 1350W [71]) employing the different heating time (from 30-45s [16] to 1-10 min [71]). Unfortunately till date, there is no theoretical or systematic experimental investigations have shown how the heating rate (MW power and, heating time) influences the final results of synthesis. Some interesting results obtained in shape-controlled synthesis [16,159] can be attributed rather to using the new reduction agents like 2, 7 dihydroxy naphthalene [159] or glucose, sucrose and maltose [16], than to using MW heating.

3. MAGNETIC NANOPARTICLE SYNTHESIS

There are only three elements which are ferromagnetic at room temperature: iron, cobalt and nickel. Additionally some compounds and alloys of manganese, chromium and europium can exhibit ferromagnetic behavior [2]. But our aim is not the discussion about different kinds of magnetic particles and their properties, but the survey of the methods which are used for their synthesis. Therefore, generally, talking about magnetic nanoparticles, MNPs, we will assume these three metals, their alloys and oxides.

Synthesis of MNPs has its own characteristic features that make it different from the synthesis of noble NPs. First, these particles more strongly aggregate to each other owing to their magnetism additional to van der Waals attraction. Second problem associates with the oxidation of metallic nanoparticles. This particle instability towards oxidation increases as the particle size gets smaller. Comparing with gold and silver aqueous colloids, which can keep their properties unchangeable for months, magnetic nanoparticles can easily be oxidized not only in air, but also in water solution in the presence of dissolved oxygen. Usually, a surfactant or polymeric protection layer can help to prevent the particle oxidation (at least, partly), but such protection layer can also change the magnetic properties of the particles strongly [160-163]. Third problem is particle crystallinity. During the synthesis of noble NPs, it is very important to control the particle size and the shape, for MNPs not only size and shape but crystal structures of the particle plays a dominant role. Magnetic properties of monocrystal, polycrystal, and amorphous particle will be essentially different. The majority of the synthetic methods usually yield amorphous nanoparticles, and further high-temperature heating (annealing) is required for their crystallization. Additional particle aggregation can be caused by this annealing process. To satisfy

these contradictory demands for particle crystallinity and non agglomeration is not so easy.

We are now considering the experimental methods used for synthesis of magnetic nanoparticles in aqueous solutions. Similar to noble metals nanoparticles, the main synthetic method for colloidal magnetic nanoparticles in aqueous solution is chemical reduction. The main sources of metal ions in water solution are the chlorides, nitrates and sulfates of iron, cobalt and nickel. The range of reducing agents for magnetic particle synthesis is narrower when comparing it with the wide range of chemical substances using as the reducing agents for the synthesis of noble metal nanoparticles. The reason for this can be explain with a thermodynamic approach. In order to transfer electron from reducing agent to metal ion, the free energy change, ΔG must be favorable. As a matter of convention, the favorability of oxidation-reduction processes is reflected in standard electrode potential, E^0, of the corresponding half-reaction (reaction (1) or (2)). The E^0 value characterizes the ability of molecule (ion) to be reduced (oxidized). The bigger the E^0 value, the easier it is to be reduced. On the other hand, the more negative the E^0 value, the more stronger reducing agent this molecule (ion) is. Standard reduction potentials for ions of noble metals are much higher than ones for magnetic metals. For $[AuCl_4]^-$ ion, reduced by reaction

$$[AuCl_4]^- + 3e_{aq} \rightarrow Au(s) + 4Cl^- \qquad (4)$$

the E^0 value is +0.93V [164], and for Ag^+ ion, reduced by reaction

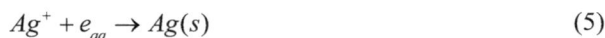

$$Ag^+ + e_{aq} \rightarrow Ag(s) \qquad (5)$$

the E^0 value is +0.80V [164, 165]. It means that both ions can easily be reduced, and very weak reducing agent, such as citrate or ascorbic acid is sufficient for their reduction. For the reduction of magnetic metals the E^0 values are much smaller [164-166]:

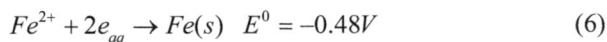

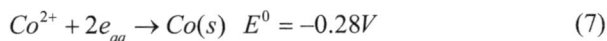

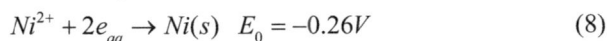

$$Fe^{2+} + 2e_{aq} \rightarrow Fe(s) \quad E^0 = -0.48V \qquad (6)$$

$$Co^{2+} + 2e_{aq} \rightarrow Co(s) \quad E^0 = -0.28V \qquad (7)$$

$$Ni^{2+} + 2e_{aq} \rightarrow Ni(s) \quad E_0 = -0.26V \qquad (8)$$

It means that the ions of magnetic metals need more stronger reducing agents for their reduction. Actually, in the majority of works, where the synthesis of magnetic nanoparticles was made by chemical reduction in aqueous solutions, borohydride or hydrazine is used as a reducing agent. Moreover, hydrazine has a weak reducing ability when the PH is low, usually when hydrazine used as reducing agent, NaOH added to water colloid to adjust suitable PH value. Nevertheless, some researchers use several other reducing agents: urea [167-168], ethanol [169], polyetilenglicol [170], PVP [171], NH$_4$·HCO$_3$ [172], NH$_4$OH [173]. Some authors used phosphorus compounds as reducing agents [174-175], however, using phosphorus can lead to the formation of metalphosphates [175], and using of borohydride sometimes can lead to formation of metalborates [176].

Almost same stabilizers, which were used for the synthesis of noble metal NPs, are being used for the protection of magnetic NPs: citrate [173, 177-180], PEG [176], PVP [181], different surfactants such as CTAB [182-185] or SDS [186], as well as polymers such as PAA [187-189] or PMAA [190-191]. As it was mentioned above, stabilizer cannot always protect a particle from oxidation. For example, XRD analysis in work [176] showed that in spite of using PEG stabilizing layer, iron particles were partly oxidized. When different researchers used just the same reducing agent and stabilizer, the results can be different. For example, by using hydrazine and CTAB for the synthesis of Ni NPs, different researchers reported different particle sizes: from 50 to 250 nm [185], from 7 to 20 nm [183], from 10 to 36 nm [184]. Authors [179] showed that the results depend on the concentrations of reagents. Using borohydride and citrate as stabilizers they produced cobalt NPs with different particle size distribution, depending on the concentration of citrate.

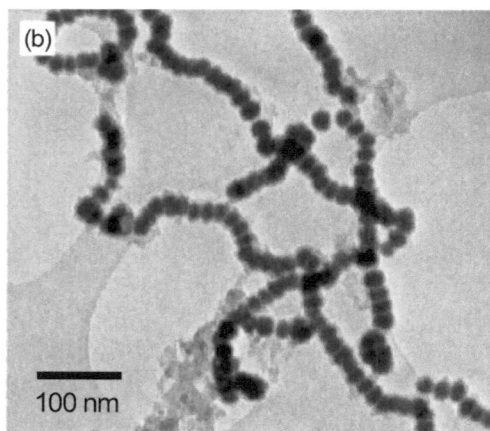

Fig. (7). TEM image of encapsulated nanoparticle chain. Images reprinted with permission from Ref. [179] Copyright © 2006 Elsevier.

Due to the strong ability of magnetic nanoparticles for oxidation, particle coating technique was applied very often for these materials. Such thin layer of inert material was to prevent particle from oxidation in colloid, as well as from their further agglomeration. Silica, SiO$_2$, is the most popular material for such purpose [179, 192-195]. But, in some researches

different materials were used. For example, in work [196] nickel nanoparticles were coated by gold layer, in [197] cobalt particles were coated by Fe$_2$O$_3$, and in work [173] magnetite, Fe$_3$O$_4$ particles were coated by silver. As it was mentioned above, particles of magnetic metals have a tendency for more strong aggregation than the particles of noble metals. Even after coating by SiO$_2$ or other protection layer, these particles can form chains, as it is shown in Fig. (7) [179]. Such chain formation was also observed when different stabilizers, such as citrate [177], SDS [186], PAA [188] were used.

The number of researches deal with a synthesis of different shapes of magnetic nanoparticles is not so large comparing with the same ones did for noble metals NPs. Mainly these are magnetic nanorods or nanowires [167, 172, 177-178, 182, 198]. This situation can easily be explained by different applications of these two classes of nanoparticles. For noble NPs of different shapes, their optical properties can be changed dramatically. At the same time, magnetic properties of cubes, plates, or spheres of the same sizes will be closed to each others. For the same reason the seed method was not developed so much for magnetic particles. Normally, nanorods or nanowires are synthesized by using the CTAB [182] or by precise control of citrate concentration [177]. Nevertheless, when the seed method is applied it permits to synthesize monodisperse nanoparticles [188].

Laser ablation in liquid media became very popular for the synthesis of magnetic nanoparticles. But, the absolute majority of such works accomplished in organic solvents rather than in water solutions [199,201-202]. There are two reasons for such situation. First, is an oxidation problem, which has already been discussed above. Using organic solvent it is more easily to prevent particles from oxidation. Second reason is the particle size, which can be controlled much easily in organic solvent. For example, Chen and coworkers [200] prepared magnetic colloid ablating cobalt target by 532-nm laser beam in pure water and in ethanol. In case of water, particles' sizes varied from 6 to 60 nm and average particle size was 18.7 +/- 6.4 nm. When ethanol was used, the particles' size ranged from 6 to 35 nm with average particle size of 11.8 +/- 2.7 nm. Laser irradiation can be used not only for primary synthesis, but also for coating process. Zhang and coworkers [203] at first prepared iron nanoparticles in oil phase, and gold NPs in water solution (with CTAB as a stabilizer). Then, they irradiated the mixture by 532 –nm laser beam. Gold particles absorbed this laser irradiation more efficiently than iron and as a result, decomposed. Au atoms, clusters, and small nanoparticles produced in the result of such decomposition condensed on Fe NPs, forming protection shell.

Sonochemical method and micro wave heating technique is also used very widely for the synthesis of magnetic NPs. But general situation with these techniques is similar: majority of synthesizes are made in organic solvents rather than in aqueous solutions.

4. CONCLUSION

Through the different experimental methods used for metal nanoparticle synthesis in aqueous solution, method of chemical reduction was developed first, and uptil now remains the main method for the synthesis of metal nanostructures of different sizes and shapes in water colloids. A lot of different chemical substances are used as reducing agents and particle stabilizers. Recently developed method of laser ablation in liquid became very popular. A number of interesting results were already obtained by this new method. For the future, combination of these two methods, as well as any combination of physical and chemical methods, looks very promising.

5. REFERENCES

[1]   Overbeek J. T. G. in Colloidal Dispersions; J. W. Goodwin (ed.), Royal Society of Chemistry, London, 1981, p1.
[2]   Nanoparticles; G. Schmid (ed.), WILEY-VCH Verlag GmbH & Co. KGaA, Weinheim, 2004.
[3]   Metal Nanoparticles. Synthesis, Characterization, and Applications; D. L. Feldheim, and Foss C. A. Jr. (eds.) , Marcel Dekker, Ink., New York, 2002.
[4]   Poovarodom, S.; Bass, J. D.; Hwang, S. J.; Katz, A. Langmuir, 2005, 21, 12348.
[5]   Rodriguez-Gonzalez, B.; Sanchez-Iglesias, A.; Giersig, M.; Liz-Marzan, L. M. Faraday Discussions, 2004, 125, 133.
[6]   Faraday M. Phil. Trans. Roy. Soc., 1857, 147, 145.
[7]   Turkevich, J.; Stevenson, P. C.; Hillier, J. Discuss. Faraday Soc.,1951, 11, 55.
[8]   Martin, C. R. Science, 1994, 266, 1961.
[9]   Martin, C. R. Chem. Mater., 1996, 8, 1739.
[10]   Yu, Y. .; Chang, S. .; Lee, C. Wang C.R. J. Phys. Chem. B 1997, 101, 661.
[11]   Chang, S.; Shin, C.; Chen, C.; Lai, W.; Wang C. R. Langmuir, 1999, 15, 701.
[12]   Brown, K. R.; Natan, M. J. Langmuir, 1998, 14, 726.
[13]   Murphy, C. J.; Jana, N. R. Adv. Mater. 2002, 14, 80.
[14]   Sherry, L. J.; Chang, S. H.; Schatz G. C.; Van Duyne, R. P.; Wiley, B. J.; Xia, Y. Nano Lett., 2005, 5, 2034.
[15]   Sau, T. K.; Murphy, C. J. Phylosoph. Magazine, 2007, 87, 2143.
[16]   Mallikarjuna, N. N.; Varma, R. S. Crystal Growth and Design, 2007, 7, 686.
[17]   Wiley, B. J.; Im, S. H.; Li, Z. Y.; McLellan, J.; Siekkinen, A. Xia. Y. J. Phys. Chem. B, 2006, 110, 15666.
[18]   Jena, B. K.; Raj, C. R, J. Phys. Chem. C, 2007, 111, 15146.
[19]   Lu, L.; Kobayashi, A.; Tawa, K.; Ozaki, Y, Chem. Mater., 2006, 18, 4894.
[20]   Yamamoto, M.; Kashiwagi, Y.; Sakata, T.; Mori, H.; Nakamoto, M. Chem. Mater., 2005, 17, 5391.
[21]   Xie, J.; Lee, J.; Wang, D. I. J. Phys. Chem. C, 2007, 111, 10226.
[22]   Zhao, N.; Wei, Y.; Sun, N.; Chen, Q.; Bai, J.; Zhou, L.; Qin, Y.; Li. M.; Qi, L. Langmuir, 2008, 24, 991.
[23]   Huang, T.; Cheng, T.; Ming, Y. Wei, H.; Lung, S.; Fu, R.; Ji, J.; Chi, Y.; Hsin, T. Langmuir, 2007, 23, 5722.
[24]   Lee, Y. W..; Kim, N. H.; Lee, K. Y.; Kwon, K.; Kim, M.; Han, S. W. J. Phys. Chem. C 2008, 112, 6717.
[25]   Wang, T.; Xiaoge, H.; Dong, S. J. Phys. Chem. B, 2006, 110, 16930.
[26]   Jena, B. K.; Raj, C. R. Langmuir, 2007, 23, 4064.
[27]   Creighton J. A. Metal Colloids in Surface Enhanced Raman Scattering, R. K. Chang and T. E. Furtak (Eds.), Plenum press, New York, 1982, pp.315-337.
[28]   Thiele, H.; Schroder, H.; Levern, V. J. Coll. Sci. 1965, 20, 679.
[29]   Curtis, A. C.; Duff, D. G.; Edwards, P.P.; Jefferson, D. A.; Johnson, B. F. G.; Kirkland, A. I.; Wallace, A. S. J. Phys. Chem. 1988, 92, 2270.
[30]   Angebrannt, M. J.; Winefordner, J. D. Talanta, 1992, 39, 569.
[31]   Hirai, H. J. Macromol. Sci.- Chem., 1979, A13, 633.
[32]   Ershov, B. G.; Janata, E.; Michaelis, M; Henglein, A. J. Phys. Chem. 1991, 95, 8996.
[33]   Ershov, B. G.; Janata, E.; Henglein, A. Rad. Phys. Chem., 1992, 39, 123.
[34]   Wilenzick, R. M.; Russell, D. C.; Morriss, R. H.; Marshall, S. W. J. Phys. Chem. 19637, 47, 533.
[35]   Kiwi, J.; Gratzel, M. J. Am. Chem. Soc. 1979, 101, 7214.
[36]   Furlomg, D. N.; Launikonis, A.; Sasse, W. H. F.; Sanders, J. V. J. Chem. Soc. Faraday Trans. 1984, 80,571.
[37]   Ahmadi, T. S.; Wang, Z. L.; Green, T. C.; Henglein, A. Science, 1996, 272, 1924.
[38]   Ahmadi, T. S.; Wang, Z. L.; Henglein, A.; ElSayed, M. A. Chem. Mater. 1996, 8,1161.
[39]   Mills, G.; Henglein, A. Rad. Phys. Chem. 1985, 26, 385.
[40]   Michaelis, M.; Henglein, A. J. Phys. Chem. 1992, 96, 4719.
[41]   Henglein, A. Chemical Processes in Inorganic Materials. : Metal and Semiconductor clusters and colloids. in: Material Research Soc. Symposium Proceedings, 1992, 272, 77.
[42]   Henglein, A. J. Phys. Chem. 1993, 97, 5457.
[43]   Schmid, S. Chem. Reviews, 1992, 92, 1709.
[44]   Cea, F.; Devenish, R. W.; Goulding, T.; Heaton, B. T.; Kiely, C. J.; Moiseev, I.; Smith, A. K.; Temple, J.; Vargaftik, M.; Whyman, R. in: Electron Microscopy and Analysis, 1993, 477.
[45]   Lee, P. C.; Meisel, D. J Phys. Chem. 1982, 86, 3391.
[46]   Van Hyning, D. L.; Zukoski, C. F. Langmuir, 1998, 14, 7034.
[47]   Creighton, J. A.; Blatchford, C. G.; Albrecht, M. G. J. Chem. Soc. Faraday Trans. 2 1979, 75, 790.
[48]   Caswell, K. K.; Bender, C. M. Murphy, C. J. Nano Lett. 2003, 3, 667.
[49]   Cason, J. P.; Khambaswadkar, K.; Roberts, C. B. Ind. Eng. Chem. Res. 2000, 39, 4749.
[50]   He, S.; Yao, J.; Jang, P.; Shi, D.; Zhang, H.; Xie, S.; Pang, S.; Gao, H. Langmuir, 2001, 17, 1571.
[51]   Nagao, D.; Shimazaki, Y.; Kobayashi, Y.; Konno, M. Coll. Surf. A 2006, 273, 97.
[52]   Chen, C. W.; Arai, K.; Yamamoto, K.; Serizawa, T.; Akashi, M. Macromol. Chem. Phys. 2000, 201,2811.
[53]   He, R.; Qian, X.; Yin, J.; Zhu, Z. J. Mater. Chem. 2002, 12, 3783.
[54]   Chen, C. W.; Takezako, T.; Yamamoto, K.; Serizawa, T.; Akashi, M. Coll. Surf. A 2000, 169, 107.
[55]   Chen, C. W.; Akashi, M. Langmuir, 1997, 13, 6465.
[56]   Okitsu, K.; Ashokkumar, M.; Grieser, F. J. Phys. Chem. B 2005, 109, 20673.
[57]   Chen, D. Huang, Y. J. Colloid Interface Sci. 2002, 255, 299.
[58]   Jacob, J. A.; Kapoor, S.; Biswas, N.; Mukherjee, T. Coll. Surf. A 2007, 301, 329.
[59]   Patel, K.; Kapoor, S.; Dave, D. P.; Mukherjee, T. Res. Chem. Intermed. 2006, 32, 103.
[60]   Sondi, I.; Goia, D. V.; Matijevic, E. J. Colloid Interface Sci. 2003, 260, 75.
[61]   Velikov, K. P.; Zegeres, G. E.; van Blaaderen, A. Langmuir 2003, 19, 1384.
[62]   Lee, J.; Kamada, K.; Enomoto, N.; Hojo, J. J. Colloid Interface Sci. 2007, 316, 887.
[63]   Wu, C. W.; Mosher, B. P.; Zeng, T. F. J. Nanopart. Research, 2006, 8, 965.
[64]   Krichevski, O.; Markovich G. Langmuir 2007, 23, 1496.
[65]   Gou, L.; Murphy, C. J. Chem. Mater. 2005, 17, 3668.

[66]   Xiong, Y.; Washio, I.; Chen, J.; Cai, H.; Li, Z. Xia, Y. Langmuir, 2006, 22, 8563.

[67]   Huang, Y. F.; Huang, K. M.; Chang H. T. J. Colloid Interface Sci. 2006, 301, 145.

[68]   Guo, Z.; Zhang, y.; Huang, L.; Wang, M.; Wang, J.; Sun, J.; Xu, L.; Gu, N. J. Colloid Interface Sci. 2007, 309, 518.

[69]   Gniewek, A.; Ziolkowski, J.; Trzeciak, A.; Kepinski, L. J. Catalysis 2006, 239, 272.

[70]   Leopold, N.; Lendl, B. J. Phys. Chem. B 2003, 107, 5723.

[71]   Pal, A.; Shan, S.; Devi, S. Coll. Surf. A 2007, 302, 51.

[72]   Panacek, A.; Kvitek, L.; Prucek, R.; Kplar, M.; Vecerova, R.; Pizurova, N.; Sharma, V. K.; Nevecna, T,; Zboril, R. J. Phys. Chem. B 2006, 110, 16248.

[73]   Rodriguez-Gattorno, G.; Diaz, D.; Rendon, L.; Hernandez-Segura, G. O. J. Phys. Chem. B 2002, 106, 2482.

[74]   Lin, C. S.; Khan M. R.; Lin, S. D. J. Colloid Interface Sci. 2006, 299, 678.

[75]   Pillai, Z. S.; Kamat, P. V. J. Phys. Chem. B 2004, 108, 945.

[76]   Patungwasa, W.; Hodak, J. H. Mater. Chem. Phys. 2008, 108, 45.

[77]   Tan, Y.; Dai, X.; Li, Y.; Zhu, D. J. Mater. Chem. 2003, 13, 1069.

[78]   Sun, Y.; Yin, Y.; Mayers, B. T.; Herricks T.; Xia, Y. Chem. Mat. 2002, 14, 4736.

[79]   Ma, H. Y.; Yin, B. S.; Wang, S. Y.; Pan, W.; Huang, S. X.; Chen, S. H,; Meng, F. J. Chem. Phys. Phys. Chem. 2004, 5, 68.

[80]   Li, X.; Zhang, J.; Xu, W.; Jia, H.; Wang, X.; Yang, B.; Zhao, B.; Li, B.; Ozaki, Y. Langmuir, 2003, 19, 4285.

[81]   Bakr, O. M.; Wunsch, B. H.; Stellacci, F. Chem. Mater. 2006, 18, 3297.

[82]   Sun, S.; Mendes, P.; Critchley, K.; Diegoli, S.; Hanwell, M.; Evans, S.; Legget, G.; Preece, J.; Richardson, T. Nano Letters, 2006, 6, 345.

[83]   Chen, C.; Tzeng, S.; Chen, H.; Lin, K.; Gwo, S. J. Am. Chem. Soc. 2008, 130, 824.

[84]   Prochazka, M.; Vlckova, B.; Stepanek, J.; Turpin, P. Langmuir, 2005, 21, 2956.

[85]   Doty, R. C.; Tshikhudo, t. R.; Brust, M.; Fernig, D. G. Chem. Mater. 2005, 17, 4630.

[86]   Maillard, M.; Girgio, S.; Pileni, M. J. Phys. Chem. B 2003, 107, 2466.

[87]   Zhang, J.; Han, B.; Liu, M.; Liu, D.; Dong, Z.; Liu, J.; Li, D.; Wang, J.; Dong, B.; Zhao, H.; Rong, L. J. Phys. Chem. B 2003, 107, 3679.

[88]   Chen, S.; Carroll, D. L. Nano Lett. 2002, 2, 1003.

[89]   Gole, A.; Murphy C. J. Chem. Mater. 2005, 17, 1325.

[90]   Miranda, O. R.; Ahmadi, T. S. J. Phys. Chem. B 2005, 109, 15724.

[91]   Keul, H. A.; Moller, M.; Bockstaller, M. R. Langmuir, 2007, 23, 10307.

[92]   Smith, D. K.; Korgel, B. A. Langmuir, 2008, 24, 644.

[93]   Kou, X.; Zhang, S.; Tsung, C.; Yang, Z.; Yeung, M.; Stucky G.; Sun, L.; Wang, J.; Yan, C. Chem.-A European J. 2007, 13, 2929.

[94]   Turkevich J. Colloid silver. Part 1. in: Gold. Bull. 1985, 18, 86.

[95]   Pyatenko, A.; Yamaguchi, M.; Suzuki, M. J. Phys. Chem. B 2005, 109, 21608.

[96]   Mulvaney, P.; Giersig, M.; Henglein, A. J. Phys. Chem. 1993, 97, 7061.

[97]   Henglein, A.; Giersig, M. J. Phys. Chem. 19994, 98, 6931.

[98]   Jana, N. R.; Gearheart, L.; Murphy, C. J. Chem. Mater. 2001,13, 2313.

[99]   Jana, N. R.; Gearheart, L; Murphy, C. J. Langmuir 2001, 17, 6782.

[100]   Sau, T. K.; Pal, A.; Jana, N. R.; Wang, Z. L.; Pal, T. J. Nanopart. Res. 2001, 3, 257.

[101]   Jana, N. R.; Gearheart, L.; Murphy, C. J. J. Phys. Chem. B 2001, 105, 4065.

[102]   Jana, N. R.; Gearheart, L.; Murphy, C. J. Adv. Mater. 2001, 13, 1389.

[103]   Murphy, C. J.; Sau, T. K.; Gole, A. M.; Orendorff, C. J.; Gao, J.; Gou, L.; Hunyadi, S. E.; Li, T. J. Phys. Chem. B 2005, 109, 13857.

[104]   Wiley, B.; Sun, Y.; Mayers, B.; Xia, Y. Chem. Eur. J. 2005, 11, 454.

[105]   Xia, Y.; Halas, N.J. MRS Bulletin, 2005, 30, 338.

[106]   Ducamp-Sanguesa, C.; Herrera-Urbina, R.; Figlarz, M. J. Solid State Chem. 1992, 100, 272.

[107]   Evanoff, D. D. Jr.; Chumanov G. J. Phys. Chem. B 2004, 108, 13948.

[108]   Gachard, E.; Remita, H.; Khatouri, J.; Keita, B.; Nadjo, L.; Belloni, J. New J. Chem. 1998, 22, 1257.

[109]   Henglein, A.; Giersig, M. J. Phys. Chem. B 1999, 103, 9533.

[110]   Henglein, A. J. Phys Chem. B 2000, 104, 2201.

[111]   Kapoor, S.; Salunke, H. G.; Pande, B. M.; Kulshreshtha, S. K.; Mittal, J. P. Mater. Research Bullet. 1998, 33, 1555.

[112]   Kapoor, S.; Joshi, R.; Mukherjee, T. Chem. Phys. Lett. 2002, 354, 443.

[113]   Kamat, P. V.; Flumiani, M.; Hartland, G. V. J. Phys. Chem. B 1998, 102, 3123.

[114]   Henglein, A. J. Phys. Chem. B 2000, 104, 1206.

[115]   Sudeep, P. K.; Kamat, P. V. Chem. Mater. 2005, 17, 5404.

[116]   Laser-induced Chemical Processes. Steinfeld, J. (ed.), Plenum Press, New York, 1981.

[117]   Laser Ablation in Material Processing: Fundamental and Applications. MRS Symposium Proceedings. Braren, B.; Dubowski, J. J.; Norton, D.P. (eds.), Pitsburg, MRS, 1993.

[118]   Bauerle, D. Laser Processing and Chemistry. Second ed. Springer-Verlag Berlin Heidelberg New York, 1996.

[119]   Laser Ablation and Desorption. In: Experimental Methods in the Physical Sciences. Miller, J. C.; Haglund, R. F. (eds.). Academic press, San Diego, 1998.

[120]   Neddersen, J.; Chumanov, G.; Cotton, T. M. Appl. Spectrosc. 1993, 47, 1959.

[121]   Sibbald, M. S.; Chumanov, G.; Cotton, T. M. J. Phys. Chem. 1996, 100, 4672.

[122]   Fojtic, A. Henglein, A. Ber. Bunsenges. Phys. Chem. 1993, 97, 252..

[123]   Mafune, F.; Kohno, J.; Takeda, Y.; Kondow, T.; Sawabe, H. J. Phys. Chem. B 2001, 105, 5114.

[124]   Mafune, F.; Kohno, J.; Takeda, Y.; Kondow, T.; Sawabe, H. J. Phys. Chem. B 2000, 104, 9111.

[125]   Mafune, F.; Kohno, J.; Takeda, Y.; Kondow, T.; Sawabe, H. J. Phys. Chem. B 2000, 104, 8333

[126]   Mafune, F.; Kohno, J.; Takeda, Y.; Kondow, T. J. Phys. Chem. B 2002, 106, 8555.

[127]   Mafune, F.; Kohno, J.; Takeda, Y.; Kondow, T. J. Phys. Chem. B 2002, 106, 7577.

[128]   Mafune, F.; Kohno, J.; Takeda, Y.; Kondow, T. J. Phys. Chem. B 2001, 105, 9050.

[129]   Mafune, F.; Kohno, J.; Takeda, Y.; Kondow, T. J. Phys. Chem. B 2003, 107, 4218.

[130]   Mafune, F.; Kondow, T. Chem. Phys. Lett. 2004, 383, 343.

[131]   Pyatenko, A.; Shimokawa, K.; Yamaguchi, M.; Nishimura, O.; Suzuki, M. Appl. Phys. A, 2004, 79, 803.

[132]   Bae, C. H.; Nam, S. H.; Park, S. M. Appl. Surface Sci. 2002, 197, 628.

[133]   Chen, Y. H.; Yeh, C. S. Colloid and Surf. A 2001, 197, 133.

[134]   Kabashin, A. V.; Meunier, M.; Kingston, C.; Luong, G. H. T. J. Phys. Chem. B 2003, 107, 4527.

[135]   Kabashin, A. V.; Meunier, M. J. Appl. Phys. 2003, 94, 7941.

[136]   Sylvestre, J. P.; Poulin, S.; Kabashin, A. V.; Sacher, E.; Meunier, M. Luong, G. H. T. J. Phys. Chem. B 2004, 108, 16864.

[137]   Smejkal, P.; Peleger, J.; Siskova, K.; Viskova, B.; Dammer, O.; Slouf, M. Appl. Phys. A, 2004, 79, 1307.

[138]   Bozon-Verduraz, F.; Brayner, R.; Voronov, V. V.; Kirichenko, N. A.; Simakin, A. V.; Shafeev, G. A. Quantum Electr. 2003, 33, 714.

[139]   Simakin, A. V.; Voronov, V.V.; Shafeev, G. A.; Brrayner, R.; Bozon-Verduraz, F. Chem Phys. Lett. 2001, 348, 182.

[140]   Simakin, A. V.; Voronov, V. V.; Kirichenko, N. A.; Shafeev, G. A. Appl. Phys. A, 2004, 79, 1127.

[141]   Kazakevich, P. V.; Simakin, A. V.; Voronov, V. V.; Shafeev, G. A. Appl. Surface Sci. 2005, 252, 4373.

[142] Prochazka, M.; Mojzes, P.; Stepanek, J.; Vickova, B.; Turpin, P. Anal. Chem. 1997, 69, 5103.

[143] Izgaliev, A. T.; Simakin, A. V.; Shafeev, G. A.; Bozon-Verduraz, F. Chem Phys. Lett. 2004, 390, 467.

[144] Ding, L.; Guo, H.; Zhang, J.; Zhang, Y.; He, T.; Mo, Y. Spectroscopy and Spectr. Analys. 2008, 28, 2053.

[145] Pyatenko, A.; Yamaguchi, M.; Suzuki, M. J. Phys. Chem. 2007, 111, 7910.

[146] Suslick, K. S. The chemistry of ultrasonic. From Yearbook of Science & the Future. Encyclopedia Britanica: Chicago, 1994, p.138.

[147] Meyers, R. A. Encycl. Phys. Sci. Technol. 2001, 3, 361.

[148] Nagata, Y.; Watanabe, Y.; Fujita, S.; Dohmaru, T.; Taniguchi, S. J. Chem. Soc., Chem. Commun. 1992, 1620.

[149] Yeung, S. A.; Hobson, R.; Biggs, S.; Grieser, F. J. Chem. Soc., Chem. Commun. 1993, 378.

[150] Okitsu, K.; Bandow, H.; Maeda, Y,; Nagata, Y. Chem. Mater. 1996, 8, 315.

[151] Okitsu, K.; Ashokkumar, M.; Grieser, F. J. Phys. Chem. B 2005, 109, 20673.

[152] Haas, I.; Shanmugam, S.; Gedanken, A. J. Phys. Chem. B 2006, 110, 16947.

[153] Nemamcha, A.; Rehspringer, J.; Khatmi, D. J. Phys, Chem. B 2006, 110, 383.

[154] Radziuk, D.; Shchukin, D.; Mohwald, H. J. Phys. Chem. C 2008, 112, 2462.

[155] Harpeness, R.; Gedanken, A. Langmuir, 2004, 20, 3431.

[156] Pastoriza-Santos, I.; Liz-Marzan, L. Langmuir, 2002, 18, 2888.

[157] Chen, W.; Zhao, J.; Lee, J.; Lui, Z. Mater. Chem. Phys. 2005, 91, 124.

[158] Luo, Y. Mater. Lett. 2007, 61, 1873.

[159] Kundu, S.; Peng, L.; Liang, H. Inorg. Chem. 2008, 47, 6344.

[160] Wang, L.; Luo, J.; Fan, Q.; Suzuki, M.; Suzuki, I.; Engelhard, M. H.; Lin, Y.; Kim, N.; Wang, J.; Zhong, C. J. J. Phys. Chem. B 2005, 109, 21593.

[161] Russier, V.; Petit, C.; Legrand, J.; Pileni, M. P. Phys. Rev. B 2002, 62, 3910.

[162] van Leeuwen, D. A.; van Ruitenbeek, J. M.; de Jongh, L. J.; Ceriotti, A.; Pacchioni, G. Phys. Rev. Lett. 1994, 73, 1432.

[163] Bodker, F.; Morup, S.; Linderoth, S. Phys. Rev. Lett. 1994, 73, 1432.

[164] Bard, A. J.; Parsons, R.; Jordan, J., Eds. Standard Potentials in Aqueous Solution; Marcel Dekker, Inc.: New York, 1985.

[165] Vanysek, P. Electrochemical Series. In: CRC Handbook of Chemical and Physics, 88th ed., Chemical rubber Company, 2007.

[166] Atkins, P. Physical Chemistry, 6th edition, W.H. Freeman and Company, New York, 1997.

[167] Wu, L,; Wu, Y.; Wei, H.; Shi, Y.; Hu, C. Mater. Lett. 2004, 58, 2700.

[168] Jeevanandam, P.; Koltypin, Y.; Gedanken, A. Mater. Sci. Engineer. B 2002, 90, 125.

[169] Dang, F.; Enomoto, N.; Hojo, J.; Enpuku, K. Chem. Lett. 2008, 37, 530.

[170] Wei, X.; Zhu, G.; Liu, Y.; Ni, Y.; Song, Y.; Xu, Z. Chem. Mater. 2008, 20, 6248.

[171] Liu, D.; Ren, S.; Wu, H.; Zhang, Q.; Wen, L. J. Mater. Sci. 2008, 43, 1974.

[172] Xiang, L.; Deng, X.; Jin, Y. Scripta Materialia 2002, 47, 219.

[173] Lu, Q.; Yao, K.; Xi, D.; Liu, Z.; Ning, Q. J. Magn. Magn. Mater. 2006, 301, 44.

[174] Mohapatra, S.; Pramanik, N.; Ghosh, S. K.; Pramanik, P. J. Nanosci. Nanotech. 2006, 6, 823.

[175] Xie, S.; Qiao, M.; Zhou, W.; Luo, G.; He, H.; Fan, K.; Zhao, T.; Yuan, W. J. Phys. Chem. B 2005, 109, 24361.

[176] Bonder, M. J.; Zhang, Y.; Kiick, K. L.; Papaefthymiou, V.; Hadjipanayis, G. C. J. Magn. Magn. Mater. 2007, 311,658.

[177] Ni, X.; Zhang, J.; Zhang, Y.; Zheng, H. J. Colloid Interface. Sci. 2007, 307,, 554.

[178] Ni, X.; Zhao, Q.; Zhang, Y.; Zheng, H. Europ. J. Inorg. Chem. 2007, 3, 422.

[179] Eggeman, A. S.; Petford-Long, A. K.; Dobson, P. J.; Wiggins, J.; Bromwich, T.; Dunin-Borkowski, R.; Kasama, Y. J. Magn. Magn. Mater. 2006, 301, 336.

[180] Kobayashi, Y.; Kakinuma, H.; Nagao, D.; Ando, Y.; Miyazaki, T.; Konno, M. J. Sol-Gel Sci. Tech. 2008, 47, 16.

[181] Ma, F.; Li, Q.; Huang, J.; Li, J.; J. Crystal Growth 2008, 310, 3522.

[182] Zhang, D.; Ni, X.; Zheng, H. J. Colloid Interface Sci. 2005, 292, 410.

[183] Wu, S. H.; Chen, D. H. Chem. Lett. 2004, 33, 406.

[184] Chen, D. H.; Hsieh, C. H. J. Mater. Chem. 2002, 12, 2412.

[185] Abdel-Aal, E. A.; Malekzadeh, S. M.; Rashad, M. M.; El-Midany, A. A.; El-Shall, H. Powder Tech. 2007, 171, 63.

[186] Huang, J.; He, L.; Leng, Y.; Zhang, W.; Li, X.; Chen, C.; Liu, Y. Nanotech. 2007, 18, 415603.

[187] Yang, G. C. C.; Tu, H. C.; Huang, C. H. Sepaat. Purificat. Tech. 2007, 58, 166.

[188] Huang, K. C.; Ehrman S. H. Langmuir, 2007, 23, 1419.

[189] Chou, K. S.; Huang, K. C. J. Nanopart. Research, 2001, 3, 127.

[190] Guo, Z. H.; Henry, L. L.; Palshin, V.; Podlaha, E. J. J. Mater. Chem. 2006, 16, 1772.

[191] Tural, B.; Kaya, M.; Ozkan, N.; Volkan, M. J. Nanosci. Nanotech. 2008, 8,695.

[192] Salgueirini-Maceira, V.; Correa-Duarte, M. A. J. Mater. Chem. 2006, 16, 3593.

[193] Kobayashi, Y.; Horie, M.; Konno, M.; Rodriguez-Gonzalez, B.; Liz-Marzan, L. M. J. Phys. Chem. B 2003, 107, 7420.

[194] Haddad, P. S.; Duarte, E. L.; Baptista, M. S.; Goya, G. F.; Leite, C. A. P.; Itri, R. Surface and Colloid Sci. 2004, 128, 232.

[195] Morel, A. L.; Nikitenko, S. I.; Gionnet, K.; Wattiaux, A.; Lai-Kee-Him, J.; Labrugere, C.; Chevelier, B.; Deleris, G.; Petibois, C.; Brisson, A.; Simonoff, M. ACS NANO, 2008, 2, 847.

[196] Roy, A.; Srinivas, V.; Ram, S.; Rao T. V. C. J. Phys.-Condensed Matter 2007, 19, 346220.

[197] Chakrabarti, S.; Mandal, S. K.; Chaudhuri, S. Nanotech. 2005, 16, 506.

[198] Lu, L.; Ai, Z.; Li, J.; Zheng, Z.; Li, Q.; Zhang, L. Crystal Growth & Design 2007, 7, 459.

[199] Ishikawa, Y.; Kawaguchi, K.; Shimizu, Y.; Sasaki, T.; Koshizaki, N. Chem. Phys. Lett. 2006, 428, 426.

[200] Chen, G. H.; Hong, M. H.; Lan, B.; Wang, Z. B.; Lu, Y. F.; Chong, T. C. Appl. Surf. Sci. 2004, 228, 169.

[201] Mahfouz, R.; Cadete Santos, A.; Brenier, A.; Jacquier, B.; Bertolini, J. C. Appl. Surf. Sci. 2008, 254, 5181.

[202] Kim, S.; Yoo, B. K.; Chun, K.; Kang, W.; Choo, J.; Gong, M. S.; Joo, S. W. J. Molecul. Catal. A, 2005, 226, 231.

[203] Zhang, J.; Post, M.; Veres, T.; Jakubek, Z. J.; Guan, J.; Wang, D.; Normandin, F.; Deslandes, Y.; Simard, B. J. Phys. Chem. B 2006, 110, 7122.

CHAPTER 2

SURFACTANT BASED SYNTHESIS OF METAL NANOSYSTEMS

Jian Zhang and Hanying Zhao[*]

Key Laboratory of Functional Polymer Materials, Ministry of Education, Department of Chemistry, Nankai University, Tianjin, 300071, P. R. China

Address correspondence to: Hanying Zhao, Key Laboratory of Functional Polymer Materials, Ministry of Education, Department of Chemistry, Nankai University, Tianjin, 300071, P. R. China; Email: hyzhaob@yahoo.com

Abstract: Design of consistant and facile methods for the synthesis of silver nanoparticles is a significant forward direction in the field of application of materials science, nanotechnology and medicine. A number of methods have been suggested in the literature for the synthesis of non-agglomerated nanoparticles. These include natural and synthetic polymers, biological macromolecules, latex particles, mesoporous inorganic materials, dendrimers, microgels or hydrogels, colloidal systems and others. In this chapter, we review the latest development of various surfactant based synthesis of metal nanoparticles with a few illustrations on how the stability, morphology and complexity of nanosystems differs with current methodologies.

Key words: Nanotechnology, nanoparticles, nanocluster, surfactant.

1. INTRODUCTION

The research of nanoparticles has received much attention in the fields of materials science and colloid science in recent years due to the unique size-dependent properties of nanoparticles and applications in many fields. Metallic nanoparticles, in particular, have a variety of potential applications in optics, electronics, catalysis, diagnosis, targeted biological labeling, drug delivery, and many other fields [1-6]. During the past two decades, many methods have been developed to synthesize the metallic nanoparticles.

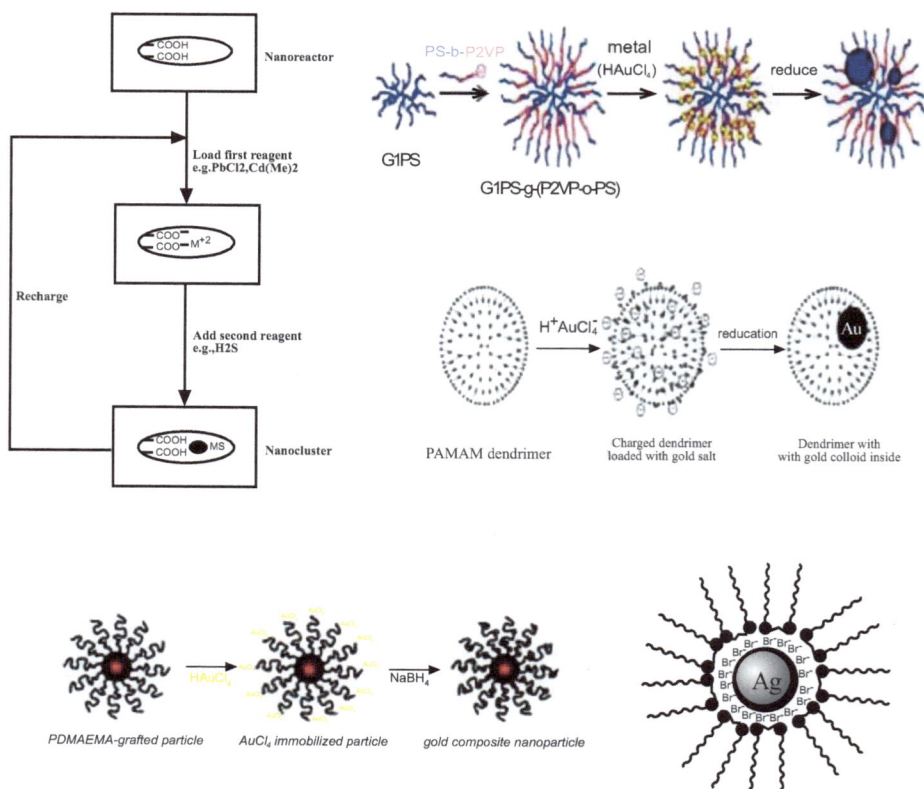

Fig. (1). Schematic illustration of some methods to synthesize the metallic nanoparticles [2-6].

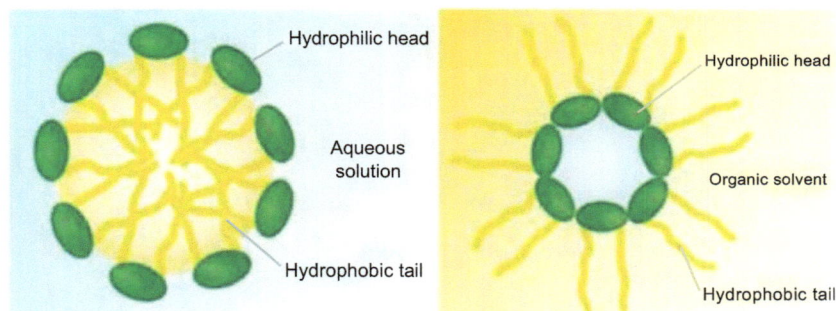

Fig. (2). Scheme representation of micelle (left) and reverse micelle (right) [www.wikipedia.org].

Most of them are based on template method, the metal precursors are limited in nano-sized domains and nanoparticles are synthesized by in situ reactions. The nano-sized domains can be treated as nanoreactors. Block copolymer films, [2] the starburst dendrimers [3] and charged polymer brushes [4] could be used as templates for the preparation of nanoparticles Fig. (1). The size and shape of the nanoparticles can be controlled by the dimensions of the nanoreactors. Synthesis of nanoparticles based on various surfactants is another important method in nanoscience. In this chapter a brief introduction to the surfactant based synthesis of metal nanosystems was made.

2. SYNTHESIS OF METAL NANOPARTICLES BASED ON SURFACTANTS

A microemulsion is obtained by shearing a mixture comprising two immiscible liquid phases with one surfactant and one co-surfactant (typical examples are hexadecane or cetyl alcohol). In an ideal microemulsion system, coalescence and Ostwald ripening are suppressed due to the presence of the surfactant and co-surfactant. Stable droplets are obtained, which have typical sizes between 50 and 500 nm. The microemulsion process is therefore particularly adapted for the generation of nanomaterials.

A micelle is an aggregate of surfactant molecules dispersed in a liquid colloid. A typical micelle in an aqueous solution is an aggregate of surfactant molecules with the hydrophilic "head" regions in contact with surrounding solvent and sequestering the hydrophobic tail in the micelle centre. In a non-polar solvent, inverse micelles are formed in which the headgroups are at the centre and the tails extending out Fig. (2). As natural soft templates, micells could provide a constrained environment during the nanoparticle growth. The size and shape of the nanoparticles could be controlled by the micelles.

Microemulsions, and in particular water-in-oil (w/o) microemulsions, have already been extensively applied in the field of metal nanoparticle synthesis where micelles, and in particular reverse micelles have been successfully used as the nanoreactors [7, 8].

For example, Pileni and cowprkers reported use of sodium bis(2-ethyl-1-hexyl) sulfosuccinate (AOT) reverse micelles system for the synthesis of silver nanoparticles [9]. O'Connor and coworkers used hexadecyltrimethyl ammonium bromide (CTAB) reverse micelles system for preparing gold nanoparticles by reduction of $HAuCl_4$ [10]. The nanocrystalline spinel magnetic nanoparticles with a narrow size distribution were prepared using precipitation in a water/CTAB, 1-butanol/1-hexanol reverse microemulsion [11, 12]. Koutzarova and co-workers prepared iron oxide (Fe_3O_4) nanoparticles in a reverse microemulsion system. The water/surfactant ratio and the metallic ion concentration exert influences on the particle size and crystallinity of nanoparticles [13].

AOT and CTAB are the most widely used surfactants. Moreover, Bumajdad and coworkers proved that cationic-non-ionic surfactant mixtures, which not only enhance aqueous phase solubility but also lead to excellent thermal stability, could be used for generation of metal oxide nanoparticles [14]. In a paper reported by Liu and co-workers silver nanoparticles were synthesized in a water-in-oil reverse micelles system, in which cationic gemini surfactant 2-hydroxy-1,3-bis(octadecyldimethyl-ammonium) propane dibromide was used as stabilizer [6].

There is considerable interest in employing surfactant-based systems as constrained reaction environments, the structure of which (spheres, rods, disks, or bicontinuous) is largely determined by the specific surfactant and their composition. There are many examples in the literature how the size and shape of the synthesized materials are correlated with the size and shape of the organized assemblies from which they originate. In a research paper, zheng and coworkers reported preparation of nickel nanoneedles in a microemulsion system using CTAB as surfactant Fig. (3). It is well known that the basic function of microemulsion droplets is to provide a confined location for the formation of nanoparticles. The weight ratio of n-octane/CTAB was adjusted to form stable threadlike micelles in the microemulsion and to prepare nanoneedles [15].

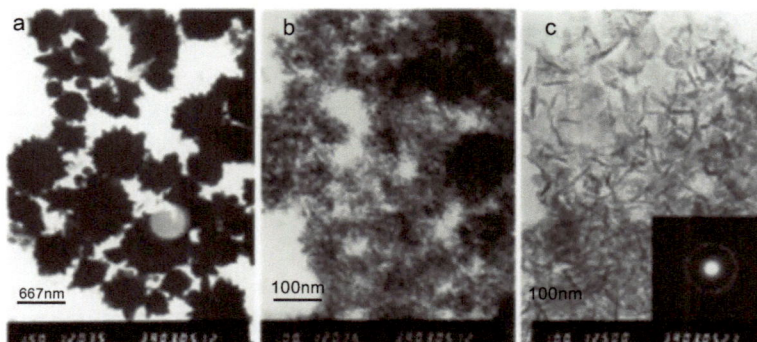

Fig. (3). Transmission electronic microscopy (TEM) images of the sample prepared with different weight ratio of CTAB/octane: (a) 0, (b) 2, (c) 20 [15].

In addition to the conventional small molecular surfactants, block copolymers can also be employed in the preparation of microemulsions and nanomaterials [16]. The micelle encapsulation approach is based solely on noncovalent interactions between nanoparticles and copolymer micelle blocks, Taton and coworkers reports an interesting approach to multifunctional hybrid nanostructures based on co-encapsulation of multiple types of nanoparticles within a block coplymer micelles [17]. This approach provides a method to prepare micelles with multiple types of nanoparticles, and with predictable physical properties Fig. (4).

Metal-organic composite is an interesting class of hybrid materials. They have shown many potential applications, such as nonlinear optics, gas adsorption, catalysis, and controlled drug release. Reverse microemulsion procedure can be used in the synthesis of nanoscale metal–organic composites. Lin and coworkers reported the surfactant-assisted synthesis of two novel gadolinium nanoscale metal–organic frameworks at elevated temperatures. Their studies showed that the two different particles obtained by

using identical building blocks are pH-dependent but temperature-dependent [18].

Liveri and coworkers prepared gold nanoparticle/ surfactant liquid crystal composites by simple evaporation of the organic solvent under vacuum. The small-size and stable gold nanoparticles were prepared by reduction of $HAuCl_4$ in the confined space of dry reverse micelles [19]. Polymerizable surfactants, for instance, cetyl-p-vinylbenzyl-dimethylammonium chloride, N,N-didodecyl-N-methyl-N-((2-methacryloyl-oxy)ethyl) ammonium chloride, and didecyldime-thylammonium methacrylate, are more and more widely used to form the reverse micellar system for subsequent nanoparticle preparation and immobilization [20].

However, there are several disadvantages in the synthesis of metal nanoparticles by using microemulsion. In order to solve these problems, some improved methods were developed in these years. One problem appearing in traditional microemulsion is separation and removal of solvent from products. In recent years, water-in-CO_2 microemulsions have been

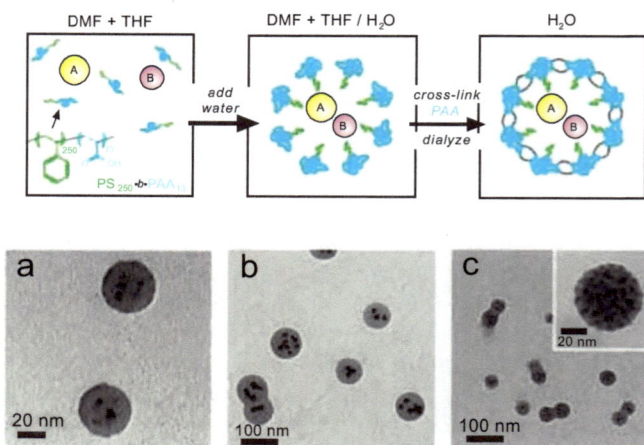

Fig. (4). Scheme preparation of multifunctional hybrid nanostructures by co-encapsulation of multiple types of nanoparticles (A and B) within a cross-linkable block copolymer micelle and TEM micrographs of multicomponent hybrid micelles prepared from combinations of different nanoparticles: (a) [CdSe/ZnS-CdSe/ZnS]@polystyrene (PS)-b-poly (acrylic acid) (PAA)$_{XL}$; (b) [Au-Fe$_3$O$_4$]@PS-b-PAA$_{XL}$; (c) [Fe$_3$O$_4$-CdSe/ZnS]@PS-b-PAA$_{XL}$; and (inset) densely packed [Fe$_3$O$_4$-CdSe/ZnS]@PS-b-PAA$_{XL}$ prepared at high particle concentration [17].

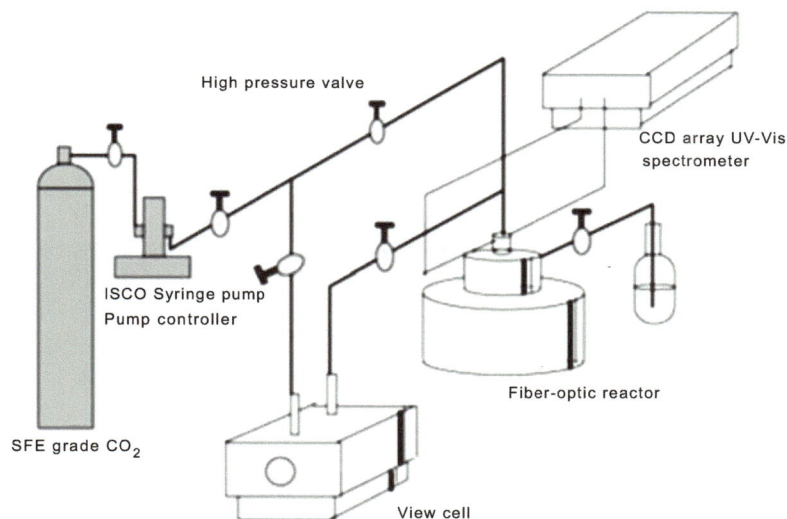

Fig. (6). Schematic diagram of the supercritical CO_2 microemulsion reaction system [21].

investigated as a reaction system for the synthesis of metal nanoparticles (Fig **6**) [21]. Synthesis of semiconductor nanoparticles in supercritical CO_2 offers several advantages over the conventional water-in-oil microemulsion approach including fast reaction speed, rapid separation and easy removal of solvent from nanoparticles. Another problem in the synthesis of metal nanoparticles by using traditional microemulsion is that the synthesized nanoparticles are usually soluble in either the aqueous phase or organic phase but not in both. But the stabilization of the the nanoparticle in different solvents is of paramount importance for their utilization as a basic unit from both fundamental and applied considerations. Many researches have been done to develop nonaqueous microemulsion for the synthesis of metal nanoparticles, because most of the organic reactions take place in nonaqueous solvents and the nanoparticles were involved in the catalysis of the reactions. Sarkar and co-workers reported in situ synthesis of siliver nanoparticles in the polar core of nonaqueous methanol reverse micelles by using $NaBH_4$ as a reducing agent and it confirms the existence of really stable reverse micelles in AOT/methanol/*n*-heptane system. This method can be applied in the synthesis of nanoparticles in other nonaqueous polar solvents with reverse micelles [22].

Adsorption and self-assembly are important features of surfactant molecules. The adsorption of surfactants from solutions onto solid surfaces has been extensively studied over the last ten years. Another kind of reactors, which are formed by surfactants self-assembled at solid /solution interfaces, also can be used as a template for the synthesis of metal nanoparticles and metal–organic nanostructure expediently. It has been reported that the apparent structure of some systems induces a templating effect that is expressed at a dimensional level several orders of magnitude larger than that of the microemulsion.

There are several reports for this synthesis approach: Kawasaki and co-workers successfully prepared platinum and gold nanoparticles by employing this kind of reactors (Fig. **7**). In their approach the self-organized surfactant nanofibers containing metal ions on a highly oriented pyrolytic graphite surface are fabricated from mixed solutions of surfactants and metal salt. This layer consists of surfactant molecules that lie horizontally on the graphite surface, exposing one side to the aqueous solution. This stripe layer acts as a template for further adsorption and finally leads to hemicylindrical micelles on the graphite. When surfactant nanofibers containing metal ions were

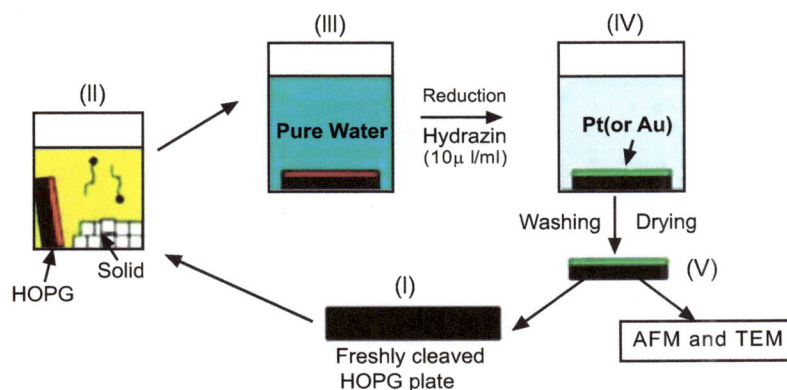

Fig. (7). Scheme preparation of synthesis of Pt nanoparticles on the nanofiber surface in repeating fashion [23].

treated with reducing agents, metal nanoparticles concentrated inside the nanofibers formed on the graphite surface [23].

Synthesis of noble metal nanoparticles via the chemical reduction of water-insoluble precursors in a mixed monolayer composed of precursor and surfactant molecules at the gas/aqueous borohydride solution interface has been carried out successfully by Khomutov and co-workers. The key to the approach is to fabricate the mixed amphiphile Langmuir monolayer by surfactant assembly at the gas/liquid interface. Time and surfactant-dependent transformations of nanoparticle morphologies and formation of 2-D aggregates were also observed. The size and shape of generated nanoparticles, and the structure of 2-D aggregates were dependent on the monolayer composition and state during the growth process [24].

Chen and co-workers reported a novel method to synthesize the anisotropic noble metal nanosystem using a biological liquid crystal as a template. Gold nanoplates mixed with spherical nanoparticles were prepared in lecithin lamellar liquid crystals containing non-ionic surfactants as reductant. The non-ionic surfactant molecules are thought to be arranged regularly in the lecithin layer separated by water to form the reactors and then $HAuCl_4$ aqueous solution is added into these systems instead of water, the $AuCl_4$ ions were reduced to the mixtures of spherical and 2-D plate-like nanoparticles by the non-ionic surfactant molecules. This implies that the soft template structure is very helpful for anisotropic growth although the products cannot replicate such template structure completely [25].

In some cases surfactants are involved in the synthesis of nanoparticles but no nanoreactors are formed in the system. In a recent work, Wang and co-workers report a facile one-step template-free surfactant-assistant solvothermal route for the synthesis of high-quality

small molecular surfactant used as the protecting agent to prepare the metal nanoparticles is CTAB, which was choosed to make cubic cuprous oxide nanocubes by reducing Cu^{2+} with sodium ascorbate. The size of these particles can be tuned by varying the concentration of surfactant Fig.(**8**) [27].

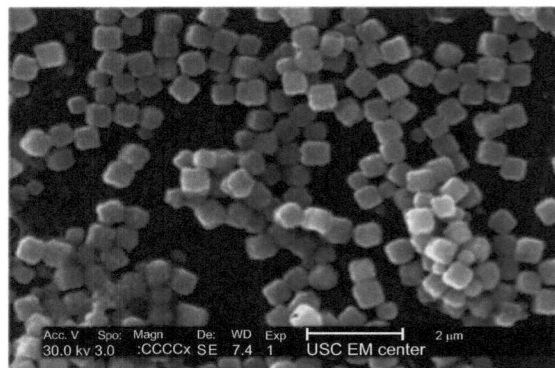

Fig. (8). Scanning electron microscopy image of cubic cuprous oxide nanocubes, coated with gold. Scale bar is 2 microns [27].

The mechanism of the surfactant-assisted system without the presence of nanoreactor was further studied by Hyeon and coworkers by using. Pd/Phosphine system [28]. The key point is the formation of a metal-surfactant complex Fig. (**9**). The monodisperse Pd nanoparticles with particle sizes of 3.5, 5, and 7 nm have been synthesized from the thermal decomposition of Pd-surfactant complexes. The particle size of Pd nanoparticles was controlled by varying the concentration of stabilizing surfactant.

Polymers carrying functional groups also can be used as specific stabilizers for the solution synthesis of nanoparticles without forming clear nanoreactors. Similar to the small molecular surfactant, the interaction between the functional groups and the metal cores plays an important role in the synthesis of

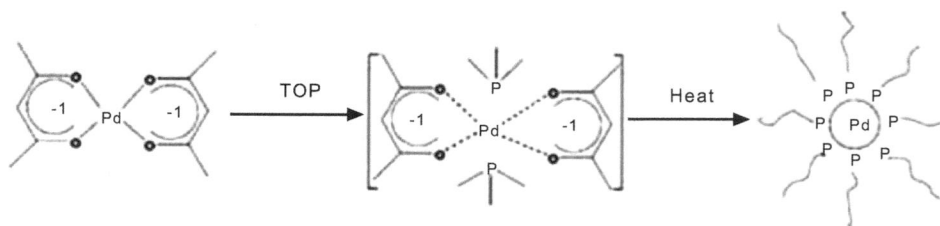

Fig. (9). Scheme representation of Pd(OAc)$_2$ precursor, Pd-trioctylphosphine complex and Pd nanoparticles [28].

Ni$_2$P hollow nanospheres using urea as the pH regulator. In the process, the surfactant sodium dodecyl sulfate (SDS) was added to prevent the aggregation of the Ni$_2$P nanospheres and used as a structure directing agent in the preparation of the nanospheres [26]. Another example for conventional

nanoparticles. Pyun and co-workers demonstrated that the end-functional polystyrene could be used to prepare polystyrene-coated ferromagnetic cobalt colloids [29]. In their research two different PS chains with either benzylamine end groups or dioctylphosphine oxide end groups were used to

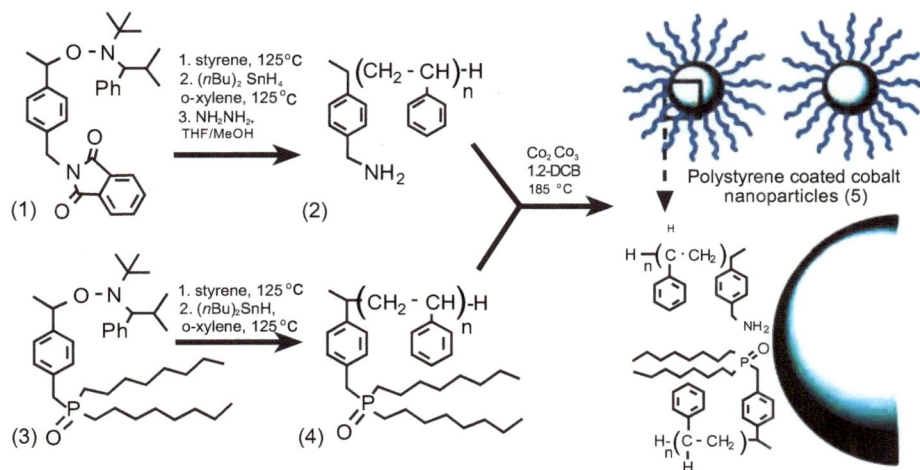

Fig. (10). Synthesis of polystyrenic surfactants and cobalt nanoparticles [29].

mimic the small molecule surfactant system Fig. (**10**). A mixture of amine terminated and phosphine oxide terminated PS chains was also used in the thermolysis of Co_2CO_8 to prepare polymer-coated cobalt nanoparticles. The combination of amine and phosphine oxide ligands was necessary to yield uniform ferromagnetic nanoparticles.

Geckeler and coworkers reported that polysorbate 80, a polymeric nonionic surfactant, can be used in the synthesis of well-dispersed gold nanoparticles in water at room temperature [30]. The synthetic procedure involves a simple one-pot process without any reducing agent Fig. (**11**). Polysorbate 80 acts both as the stabilizing agent and the reducing agent via oxidation of the oxyethylene groups into hydroperoxides, and due to its dual character, it was possible to synthesize the Au^0 particles in an aqueous solution even at a temperature of 4^oC, withour

utilizing any other agents or external energy. Thus, it is thought viable to be readily integrated into a variety of systems, especially those that are relevant to biological and biomedical applications.

Although double-hydrophilic block copolymers could not form nanoreactors (vesicles or micelles) in aqueous solutions, they also could be used as surfactants in the controlled synthesis of nanoparticles. Double-hydrophilic block copolymers consisting of a solvating poly (ethylene glycol) PEG block and a poly (ethylene imine) PEI binding block were used as effective stabilizers in the solution synthesis of high-quality CdS nanoparticles in water and methanol [31]. It was found that the chelating nitrogens of PEI block was the key to control the CdS nanoparticles and a polymer having a higher local concentration of basic nitrogens was more effective in controlling the particle size.

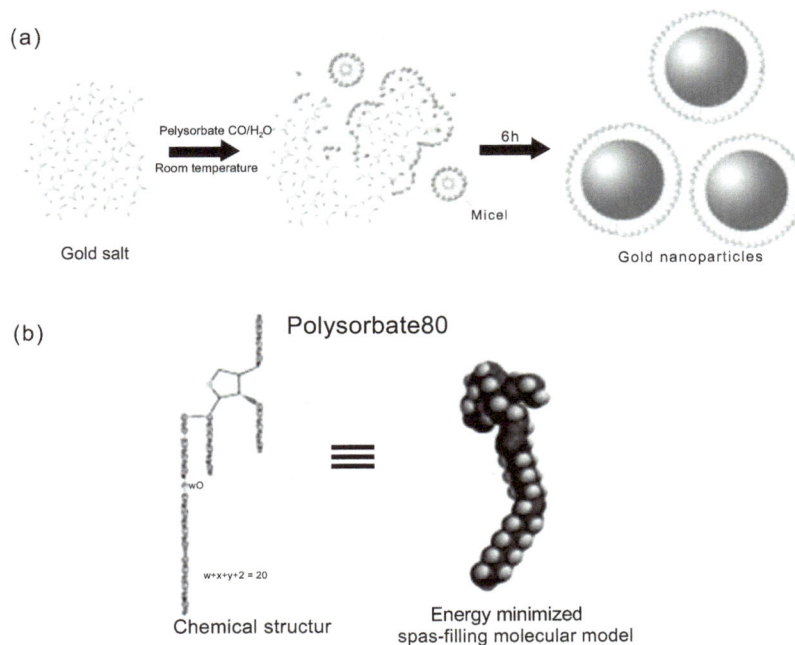

Fig. (11). (**a**) Schematic illustration of the synthesis of well-dispersed Au^0 by using polysorbate 80 at room temperature and (**b**) structure and molecular model of polysorbate 80 [30].

3. ADVANCED STRUCTURE OF METAL NANOSYSTEMS

Along with the development of the synthesis of metal nanoparticles based on surfactants, many metal nanosystems with multi-compontent and complex architectures were achieved. Many metal nanocomposites with interesting advanced structures, for instance, core/shell nanoparticles, nanorods, nanowires, nanotubes or nanosheets , had been prepared based on the pre-exist metal nanoparticles. The complexity of the surfactant behavior may be utilized to regulate the formation of structures on the micrometer scale, while their molecular structure can influence assembly processes on the nanometer scale. Base on mechanisms the preparation methods could be classified into three strategies: self-assembly process, seed-mediated process and metal nanoparticle-surfactant process.

Self-assembly approaches are becoming increasingly important, because of its potential to locate functional units at designated sites of the device. The surface features with the precursor and the spatial control are of particular importance for this method. For example,

Mathias Brust and co-workers report fabrication of gold nanowires by self-assembly of gold nanoparticles on water surfaces in the presence of surfactants. It is demonstrated that dodecanethiol-capped gold nanoparticles of 1.5-3 nm diameter in dipalmitoylphos-phatidylcholine (DPPC) self-assembled into gold nanowires Fig. (12). They believed that the unidirectional sintering of particles, which is accompanied by forming a maze-like structure, is due to a template effect of the surfactant at the molecular level. The hypothetical model is schematically illustrated in Fig. (13) [32].

Leontidis and coworkers recently reported several unusual assemblies of nanoparticles during the synthesis and subsequent dissolution of PbS particles under the action of the surfactant SDS in solutions with hydrophilic polymers [33]. In the system, PbS nanoparticles were observed to self assemble into layered superstructures, and then into nanotubes under the action of the surfactant. These nanotubes, which contain agglomerates of PbS particles and flat Pb(DS)$_2$ crystals, are similar to the "ordinary" nanotubes whose walls are not formed by nanoparticles. According to authors' claim, the key requirements for the formation

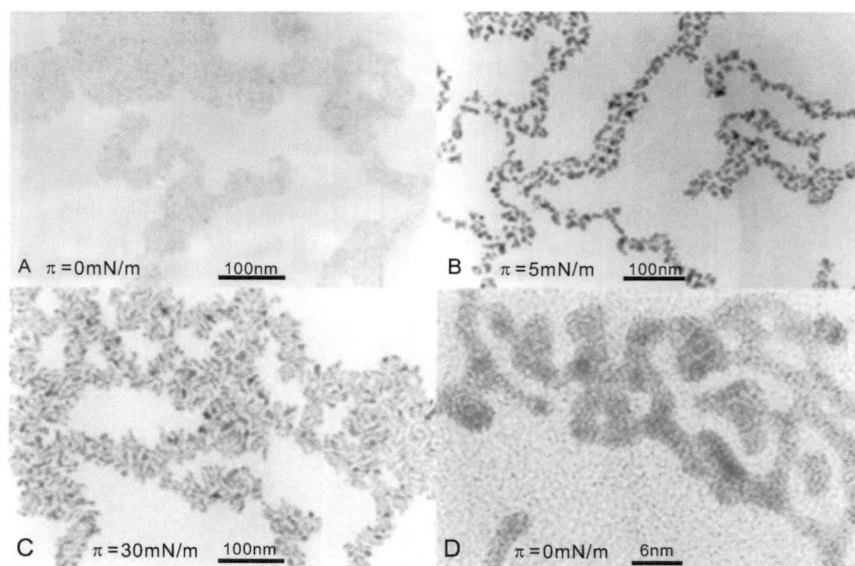

Fig. (12). TEM images of transferred Langmuir-Blodgett films onto amorphous carbon substrates at a surface pressure of **A**) 0 mN m-1, **B**) 5 mN m-1, and **C**) 30 mN m-1. **D**) High-resolution electron micrograph of the sintered gold particles seen in (C) [32].

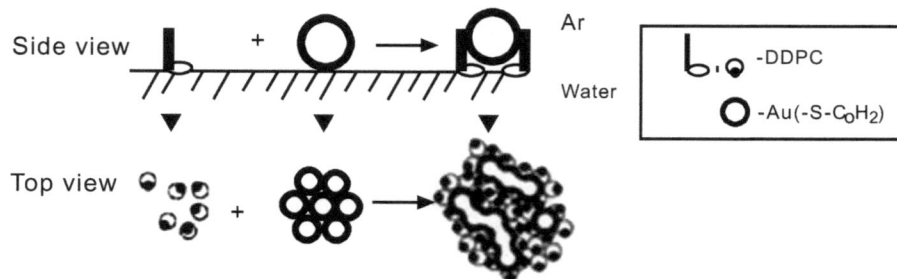

Fig. (13). A schematic model of the structures arising from the compromise between the two competing packing motifs of the DPPC and the gold particles [32].

of the nanotubes are the formation of clusters of PbS nanoparticles of roughly uniform size, and a strong association between surfactant molecules and metal ions in the system. Requirement is satisfied in polymer-surfactant systems in which the surfactant molecules associate exclusively with the polymer chains, attract Pb^{2+} ions, and create nucleation and growth domains for the subsequent reaction between Pb^{2+} and HS$^-$. The TEM images and the probable mechanism for nanotube formation are shown in Fig. (**14**) and Fig. (**15**), respectively. The process presented here show the way for the production of novel nanotube structures with the aid of appropriate surfactants. The main requirement for such processes is a strong interaction of the surfactant with the metal and the surface of the nanocrystals. However, it must be pointed out that the nanotubes prepared in this way have metastable structures [33].

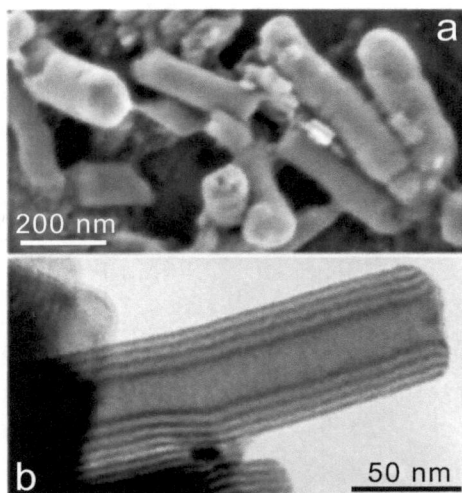

Fig. (**15**). Schematic of the proposed mechanism of nanotube formation in this system. The initial PbS precipitation in polymersurfactant rich domains is followed by adsorption of surfactant on particle surfaces, ordering and alignment of the particles by surfactant bilayers and bending of the layers to form nanotubes.The black lines represent layers containing PbS particles [33].

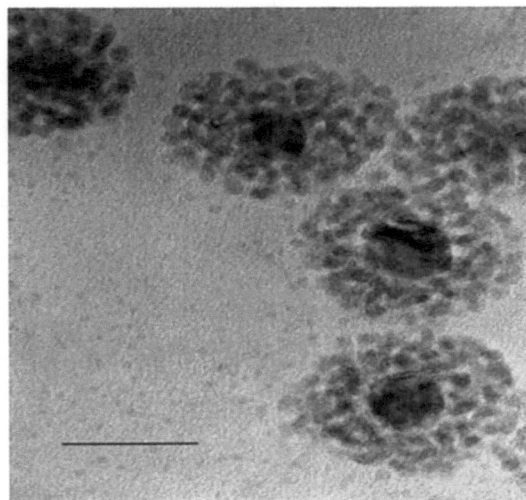

Fig. (**14**). Scanning electron microscopy (SEM) images of poly(ethylene oxide) (PEO)/SDS system (a) SEM image of a cluster of nanotubes, evolving from a region consisting of PbS particles. (b) TEM image of a well-developed nanotube [33].

Wang et al referred that in an alcohol surfactant system, a highly ordered 3D "spheres-around-sphere" type nanostructure Fig. (**16**) could be built by the assembly of the Pd nanoparticles. This configuration is also attributed to the transformation of the liquid crystal phase of the micelle molecules to a lamellar phase [34].

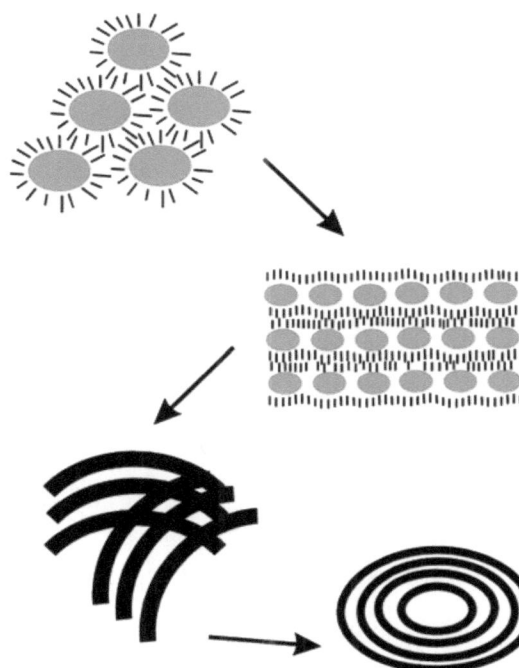

Fig. (**16**). TEM image of the "spheres-around-sphere" type Pd nanostructures. Scale bar = 20 nm [34].

Seed-mediated process is based on a seeding and fast autocatalytic growth approach in which an aqueous solution of metal salts is reduced in the presence of surfactant. Murphy's group did a lot of work on the seeding growth approach. Using a seed-mediated growth approach in a rodlike micellar media, metal nanorods were prepared from spherical nanoparticles [35, 36 and 37]. The TEM images are shown in Fig.

(17) and Fig. (18). The basic principle for this method involves two steps: the First step is the preparation of small size spherical seeds, and the second step is the growth of the seeds in rod-like micellar environment. The nanorods with high aspect ratio could be traeted as nanowires, and the aspect ratio could be controlled by properly controlling the growth conditions. Their researches proved that the only difference between the preparation of nanorods and the preparation of nanowires was the relative amount of adjusting agent in solution or the times of seed addition in successive steps. But the mechanism is not completely understood.

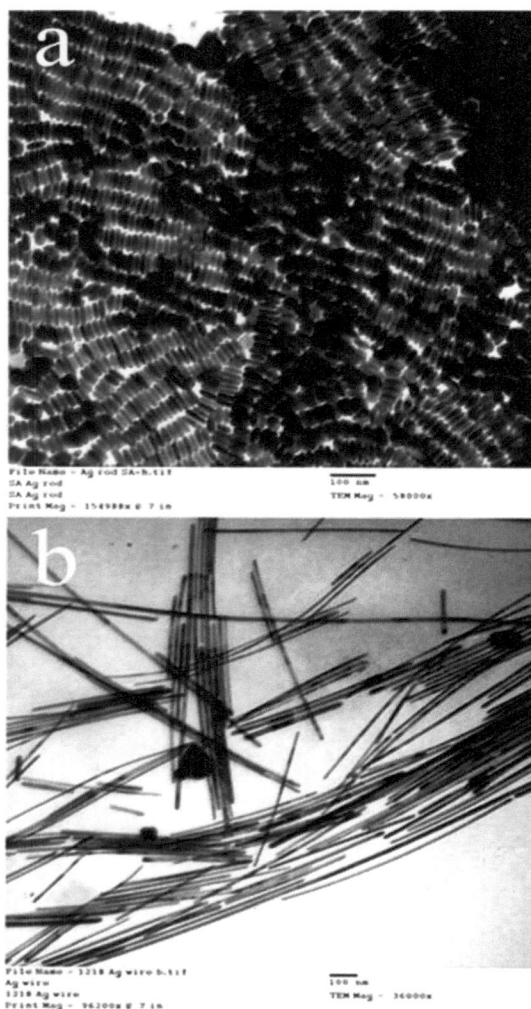

Fig. (17). TEM image of shape-separated silver nanorods (a); scale bar = 100 nm and shape-separated silver nanowires (b); scale bar = 100 nm [35].

It is very convenient to combine various different metals within a single nanostructure by seed-mediated method. Using Au nanorods as the seed, Vaia and co-workers synthesized complex Ag-tipped Au nanorods successfully, by the strong confinement of the mixed binary surfactant mixtures (CTAB/benzyldimethylhexadecylammonium chloride (BDAC)) [38].

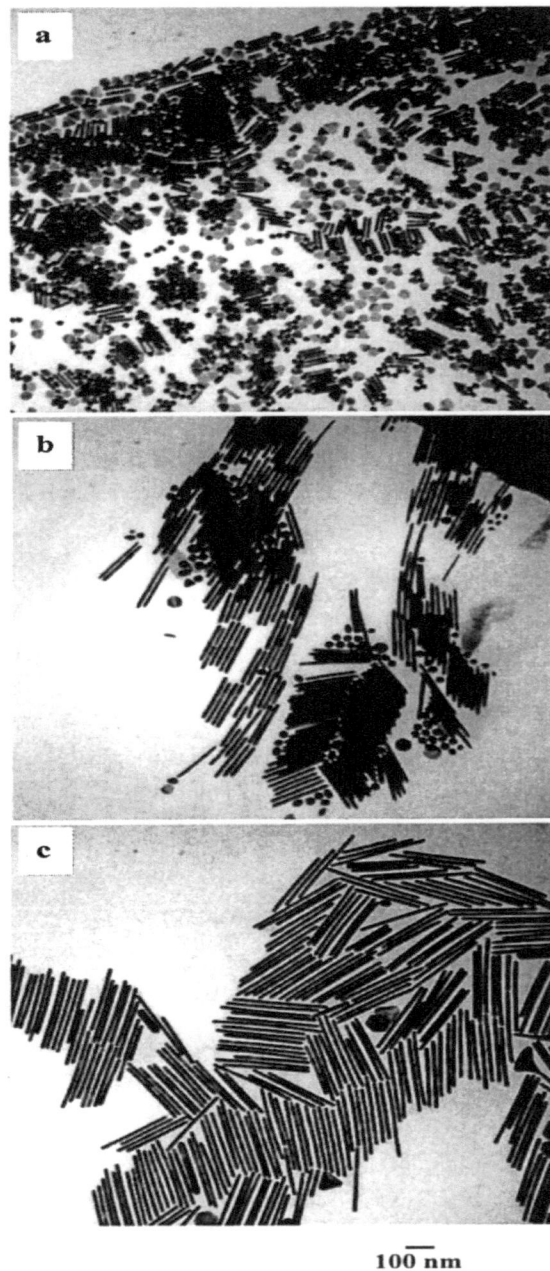

Fig. (18). (a) TEM images of 4.6 aspect ratio gold nanorods, (b) shape-separated 13 aspect ratio gold nanorods, and (c) shape-separated 18 aspect ratio gold nanorods. The scale bar (100 nm) applies to all three images [37].

Another example was reported by Xia and his co-workers [39]. The first step of their procedure is the formation of platinum nanoparticles by reducing $PtCl_2$ with ethylene glycol at about 160 °C. These platinum nanoparticles could serve as seeds for the heterogeneous nucleation and growth of silver which was conducted in the solution via the reduction of $AgNO_3$ with ethylene glycol, and uniform silver nanowires with aspect ratios as high as about 1000 were obtained in the presence of PVP. The role of PVP in this process is still not clear. One possible function for PVP was to kinetically control the growth rates of various faces by interacting with these faces through adsorption and desorption. Since the approach they used has been applied to the synthesis of a broad range of colloidal particles, it is believed that the approach could also be applied in other systems.

Metal nanoparticle-surfactant approach is a novel method to synthesize advanced hybrid metal nanocomposite. In this approach, metal nanoparticles are used to stabilize another kind of metal nanoparticles, which play the similar role to conventional surfactant molecules. The approach offers a convenient and "clear" method for producing hybrid metal nanocomposite systems with a certain control over the metal particle size without employing surfactants and/or additives.

Agostiano and co-workers used surfactant-capped anatase TiO_2 nanorods to stabilize Ag nanoparticles in nonpolar solutions in the absence of specific ligands [40]. The schematic illustration and TEM images are shown in Fig. (**19**). A fine control over the silver particle size could eventually be achieved by careful adjustment of the irradiation conditions. A high chemical reactivity characterizes the as-prepared TiO_2-nanorod-stabilized Ag nanoparticles can be potentially used for a number of applications that require "clean" metal surfaces, such as homogeneous catalysis, photocatalysis, and chemical sensing devices.

4. THE DECORATION OF METAL NANOPARTICLES BASED ON THE

INTERACTION BETWEEN SURFAC-TANT AND METAL NANOPARTICLES

Surface chemistry is of great importance to understand the chemical and physical properties of nanoparticles. It is recognized that the bare surfaces of many metallic nanoparticles are charged. However, it is very difficult to make isolated bare metal nanoparticles without surfactant molecules due to the van der Waals forces and magnetic dipolar interactions between the particles. Almost all the metal nanoparticles prepared based on surfactant have a surfactant shell outside and many properties of nanoparticles are related to the surfactant molecules. It could be illuminated clearly by mixing three types of silver nanoparticles, underivatized, surfactant-stabilized, and dodecanthiol-derivatized nanoparticles in a mesoscopically ordered lamellar gel phase of a polymer-grafted, lipid-based complex fluid [41]. It was demonstrated that silver particles capping with dodecanthiol-derivatized insert into the hydrocarbon bilayer region, while surfactant-stabilized and underivatized silver nanoparticles reside in the aqueous channels, with the latter particles preferentially interacting with the grafted PEG chains/charged membrane interface region. In the last 10 years, our knowledge of the surface modification of metal nanoparticels has improved markedly from an experimental standpoint.

One of the most important applications of surface modification is phase transfer of metal particles into an organic solvent. Because most of the chemical reactions catalysed by metal nanoparticles are carried out in organic solvent, dispersion of metal particles in

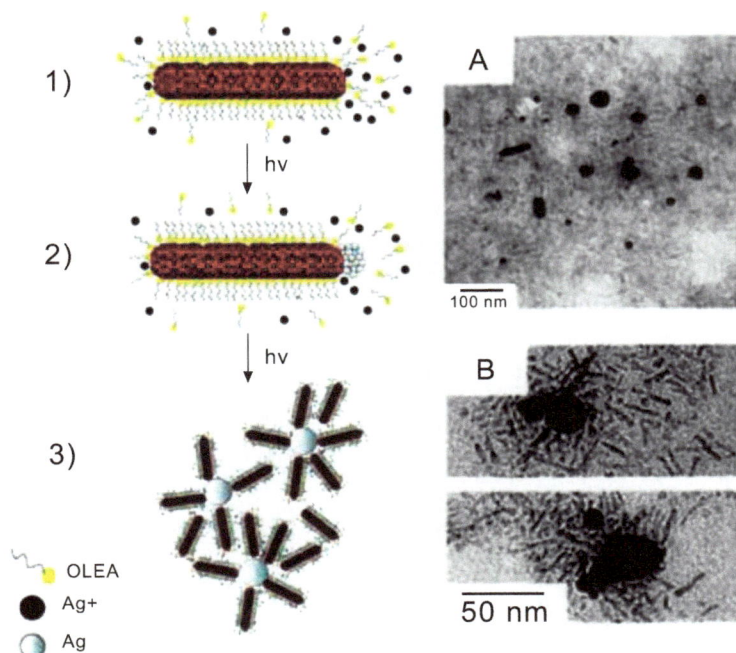

Fig. (19). Schematic of proposed mechanism for the colloidal stabilization of Ag nanoparticles by means of TiO_2 nanorods and TEM images of (A) TiO_2 nanorod/Ag nanocomposites and (B) characteristic assemblies of individual Ag nanoparticles and TiO_2 nanorods [40].

an organic solvent is necessary. Based on the means of surface modification, the nanoparticles could easily be phase transferred into organic solvents, isolated, and redispersed in a variety of organic solvents such as benzene, chloroform, and dichloromethane. Most phase-transfer experiments follow the two-phase method proposed by Brust and co-workers [42]. In the procedure, metal ions are transferred into an organic ayer by a phase-transfer reagent and then reduced in the presence of capping agents [42].

Sampath and coworkers demonstrated the phase transfer of the capped particles into toluene containing tetraoctyl ammonium bromide (TOABr) by means of an electrostatic interaction between the negatively charged mercaptopropionic acid (MPA)-capped particles and the positively charged TOABr. The appearance of the carboxylate-stretching band clearly indicates that there is an electrostatic interaction between MPA-capped particles and TOABr [43].

Surfactant exchange method can also be used in transfer of nanoparticles. Nanoparticles can be prepared in water in the presence of one surfactant and then transferred to an organic solvent with another surfactant, by which the nanoparticles are more strongly capped. Using surfactant exchange procedure, Seshadri and coworkers transfered CTAB-stabilized nanoparticles into toluene by using *n*-octylamine. The technique can easily be extended to other spinel ferrites as well and to particles of different sizes [44].

To realize their potential applications in biology and to understand the environmental implications of these nanoparticles, it is important to develop a versatile method that can be used to transfer the particles from organic phases to aqueous solutions. Yang and coworkers described a method to improve the dispersion of oleic acid stabilized nanoparticles in aqueous solutions. They modified the surface properties of the nanoparticles through the formation of an inclusion complex between surface-bounded surfactant molecules and γ-cyclodextrin (γ-CD) Fig. (**20**). Inclusion complexes between surface-bound oleic acid and CD appear to change the hydrophobic surface to a hydrophilic surface. The obtained aqueous suspensions of nanoparticles have very good stability, even at the temperature close to the boiling point of water [45].

In a homogeneous phase, the interaction of surfactant is usually employed to modify the surface of nanoparticle. Laibinis and coworkers described an effective way to synthesize stable, water-based magnetic nanoparticles stabilized by self associated bilayers of two *n*-alkanoic acids [46]. The bilayer of the primary and secondary surfactants is packed and highly organized on the iron oxide surface with some interdigitation of the two layers Fig (**21**). And it is also found that different secondary surfactants can be used

to stabilize the particles coated with the same primary surfactant to produce aqueous magnetic fluids.

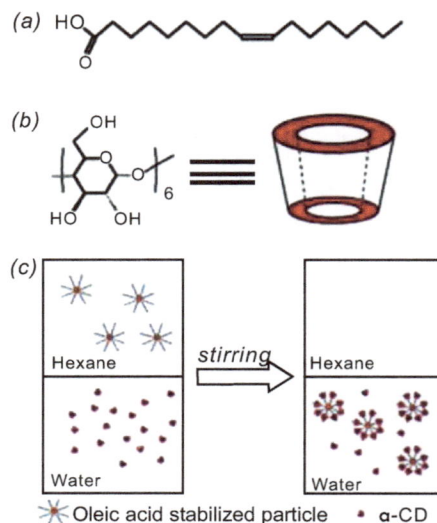

Fig. (20). Chemical structures of (a) oleic acid, (b) r-CD molecules, and (c) a schematic illustration of transfer of oleic acid stabilized nanoparticles from organic into aqueous phase by surface modification using CD [45].

Although the success of these synthesis and modification are indisputable, we are still lack of a detailed understanding of the interaction between the surfactant and the nanoparticles. Several fundamental questions still exist, including what is the atomistic structure of the center and the surface of metal nanoparticles, what is the nature of the interface between the metal dots and the surface surfactants and so on. So understanding the interaction between the surfactant and the nanoparticles is critical and essential toward the ultimate goal of designing nanostructures from the bottom up, with tailored structural [47]. Delightedly, some initial progress has been achieved. In a research reported by Dravid and coworkers cobalt nanoparticles (~15 nm in size) were synthesized with the fatty acid (oleic acid) as a surfactant, and the interaction between the surfactant and the nanoparticles was studied [48]. The Fourier transform infrared spectroscopy (FIRT) and X-ray photoelectron spectroscopy (XPS) results indicated that oleic acid gets chemisorbed as a carboxylate on the nanoparticle surfaces with the formation of the covalent bond. Because oleic acid is widely used as a surfactant in colloids synthesis, this study can help to better understand the interaction and the chemistry between the surfactant and the metal nanoparticles. In another research, Williamson and coworkers performed accurate, first-principles electronic structure simulations to study the atomistic detail of the interaction between organic surfactant molecules and the surfaces of CdSe semiconductor nanoparticles [47]. The calculated relative binding energies on different facets support the hypothesis that those facets most strongly bound to ligands are the slowest to grow and the relative binding energies control the relative growth rates of different facets, and hence the resulting geometry of the nanoparticles.

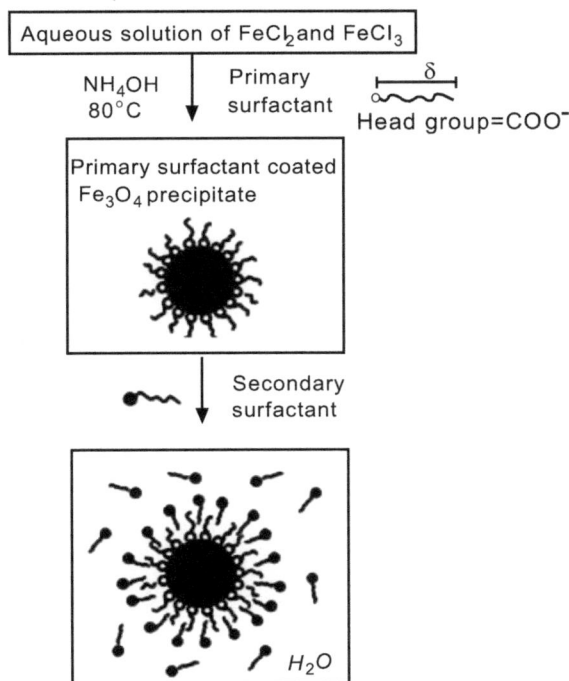

Fig. (21). Schematic representation for the synthesis of surfactant bilayer stabilized magnetic fluids, using fatty acids as the primary and secondary surfactants to produce stable aqueous magnetic fluids [46].

5. SUMMARY AND OUTLOOK

Currently, an increasing amount of work is being published on metal nanoparticles. The recent developments on the preparation of this type of nanosystem based on surfactants were reviewed.

Some methods for the preparation of metal nanoparticles and advanced structures have been investigated extensively. Besides that, the interaction between surfactants and metal nanoparticles has been introduced. Although much work has already been done on various aspects of surfactant-based metal nanoparticles, more researches are required.

By taking advantage of the various surfactants and metal ions, novel combinations of properties and more and more functional advanced structures will be achieved. Moreover, the interactions between surfactants and nanoparticles play important roles in enhancing or limiting the overall properties of the system. Some trends are observed, but no universal models for the interactions were created yet. Scientific boldness is required to search for new surfactants and illustrate the mechanism of interactions. New metal nanosystems which can be reliably utilized in real life applications are expected to be developed.

6. REFERENCES

[1] Zhao, H.; Douglas, E. P.; Harrison, B. S.; Schanze, K. S. Langmuir, 2001, 17, 8428.
[2] Kane, R. S.; Cohen, R. E.; Silbey, R. Chem. Mater., 1996, 8, 1919.
[3] (a) Gröhn, F.; Bauer, B. J.; Akpalu, Y. A.; Jackson, C. L.; Amis, E. J. Macromolecules, 2000, 33, 6042. (b) Dockendorff, J.; Gauthier, M.; Mourran, A.; Möller, M. Macromolecules, 2008, 41, 6621.
[4] Zhang, M.; Liu, L.; Wu, C.; Fu, G.; Zhao, H.; He, B. Polymer, 2007, 48, 1989.
[5] Zhao, H.; Douglas, E. P. Chem. Mater., 2002, 14, 1418.
[6] Xu, J.; Han, X.; Liu, H.; Hu, Y. Colloids Surf. A., 2006, 273, 179.
[7] Rees, G. D.; Evans-Gowing, R.; Hammond, S. J.; Robinson, B. H. Langmuir, 1999, 15, 1993.
[8] Lee, Y.; Lee, J.; Bae, C. J.; Park, J.; Hoh, H.; Park, J.; Hyeon, T. Adv. Funct. Mater., 2005, 15, 503.
[9] Taleb, A.; Petit, C.; Pileni, M. P. Chem. Mater., 1997, 9, 950.
[10] Lin, J.; Zhou, W.; O'Connor, C. J. Mater. Lett., 2001, 49, 282.
[11] Košak, A.; Makovec, D.; Drofenik, M. Phys. Stat. Sol. (c), 2004, 1, 3521.
[12] Košak, A.; Makovec, D.; Drofenik, M.; Žnidaršič, A. J. Magn. Magn. Mater., 2004, 272, 1542.
[13] Koutzarova, T.; Kolev, S.; Ghelev, Ch.; Paneva, D.; Nedkov, I. Phys. Stat. Sol. (c), 2006, 3, 1302.
[14] Bumajdad, A.; Eastoe, J.; Zaki, M. I.; Heenan, R. K.; Pasupulety, L. J. Colloid Interface Sci., 2007, 312, 68.
[15] Zhang, D. E.; Ni, X. M.; Zheng, H. G.; Li, Y.; Zhang, X. J.; Yang, Z. P. Mater. Lett., 2005, 59, 2011.
[16] (a) Gan, L. M.; Chew, C. H.; Lee, K. C.; Ng, S. C. Polymer, 1994, 35, 2659. (b) Larpent, C.; Bernard, E.; Richard, J.; Vaslin, S. React. Funct. Polym., 1997, 33, 49. (c) Petit, C.; Jain, T. K.; Billoudet, F.; Pileni, M. P. Langmuir, 1994, 10, 4446. (d) Boltri, L.; Canal, T.; Esposito, P.; Carli, F. Eur. J. Pharm. Biopharm., 1995, 41, 70. (e) Morel, S.; Ugazio, E.; Cavalli, R.; Gasco, M. R. Int. J. Pharm., 1996, 132, 259.
[17] Kim, B.; Taton, T. A. Langmuir, 2007, 23, 2198.
[18] Taylor, K. M. L.; Jin, A.; Lin, W. Angew. Chem., 2008, 47, 1.
[19] Calandra, P.; Giordano, C.; Longo, A.; Liveri, V. T. Mater. Chem. Phys., 2006, 98, 494.
[20] Hirai, T.; Watanabe, T.; Komasawa, I. J. Phys. Chem. B, 2000, 104, 8962.
[21] Ohde, H.; Ohde, M.; Bailey, F.; Kim, H.; Wai, C. M. Nano Lett., 2002, 2, 721.
[22] Setua, P.; Chakraborty, A.; Seth, D.; Bhatta, M. U.; Satyam, P. V.; Sarkar, N. J. Phys. Chem. C, 2007, 111, 3901.
[23] Kawasaki, H.; Uota, M.; Yoshimura, T.; Fujikawa, D.; Sakai, G.; Arakawa, R.; Kijima, T. Langmuir, 2007, 23, 11540.
[24] Khomutov, G. B.; Gubin, S. P. Mater. Sci. Eng. C, 2002, 22, 141.
[25] Wang, L.; Chen, X.; Sun, Z.; Chai, Y. Can. J. Chem. Eng., 2007, 85, 598.
[26] Wang, X.; Wan, F.; Gao, Y.; Liu, J.; Jiang, K. J. Crystal Growth, 2008, 310, 2569.
[27] Gou, L.; Murphy, C. J. Nano Lett., 2003, 3, 231.
[28] (a) Son, S. U.; Jang, Y.; Yoon, K. Y.; Kang, E.; Hyeon, T. Nano Lett., 2004, 4, 1147. (b) Kim, S.; Park, J.; Jang, Y.; Chung, Y.; Hwang, S.; Hyeon, T.; and Kim, Y. W. Nano Lett., 2003, 3, 1289.
[29] Korth, B. D.; Keng, P.; Shim, I.; Bowles, S. E.; Tang, C.; Kowalewski, T.; Nebesny, K. W.; Pyun, J. J. Am. Chem. Soc., 2006, 128, 6562.
[30] Premkumar, T.; Kim, D.; Lee, K.; Geckeler, K. E. Macromol. Rapid Commun., 2007, 28, 888.
[31] Qi, L.; Cölfen, H.; Antonietti, M. Nano Lett., 2001, 1, 61.
[32] Hassenkam, T.; Nørgaard, K.; Iversen, L.; Kiely, C. J.; Brust, M.; Bjørnholm, T. Adv. Mater., 2002, 14, 1126.

[33]   Leontidis, E.; Orphanou, M.; Kyprianidou-Leodidou, T.; Krumeich, F.; Caseri, W. Nano Lett., 2003, 3, 569.

[34]   Lee, C.; Wan, C.; Wang, Y. Adv. Funct. Mater., 2001, 11, 344.

[35]   Jana, N. R.; Gearheart, L.; Murphy, C. J. Chem. Commun., 2001, 7, 617

[36]   Jana, N. R.; Gearheart, L.; Murphy, C. J. Adv. Mater., 2001, 13, 1389.

[37]   Jana, N. R.; Gearheart, L.; Murphy, C. J. J. Phys. Chem. B, 2001, 105, 4065.

[38]   Park, K.; Vaia, R. A. Adv. Mater., 2008, 20, 1.

[39]   Sun, Y.; Gates, B.; Mayers, B.; Xia, Y. Nano Lett., 2002, 2, 165.

[40]   Cozzoli, P. D.; Comparelli, R.; Fanizza, E.; Curri, M. L.; Agostiano, A.; Laub, D. J. Am. Chem. Soc, 2004, 126, 3868.

[41]   Firestone, M. A.; Williams, D. E.; Seifert, S.; Csencsits, R. Nano Lett., 2001, 1, 129.

[42]   (a) Brust, M.; Walker, M.; Bethell, D.; Schiffrin, D.J.; Whyman, R. J. Chem. Soc., Chem. Commun., 1994, 7, 801. (b) Brust, M.; Fink, J.; Bethell, D.; Schiffrin, D.; Kiely, C. J. Chem. Soc.Chem. Commun., 1995, 16, 1655.

[43]   Devarajan, S.; vimalan, B.; Sampath, S. J. Colloid Interface Sci., 2004, 278, 126.

[44]   Ghosh, M.; Lawes, G.; Gayen, A.; Subbanna, G. N.; Reiff, W. M.; Subramanian, M. A.; Ramirez, A. P.; Zhang, J.; Seshadri, R. Chem. Mater., 2004, 16, 118.

[45]   Wang, Y.; Wong, J. F.; Teng, X.; Lin, X. Z.; Yang, H. Nano Lett., 2003, 3, 1555.

[46]   Shen, L.; Laibinis, P. E.; Hatton, T. A. Langmuir, 1999, 15, 447.

[47]   Puzder, A.; Williamson, A. J.; Zaitseva, N.; Galli, G.; and Manna, L.; Alivisatos, A. P. Nano Let., 2004, 4, 2361.

[48]   Wu, N.; Fu, L.; Su, M.; Aslam, M.; Wong, K. C.; Dravid, V. P. Nano Lett., 2004, 4, 383.

Recent Advances in Nanoscience and Technology, 2009, 25-37

CHAPTER 3

MICROEMULSION MEDIATED SYNTHESIS OF NANOPARTICLES

Deepa Sarkar and Kartic C. Khilar[*]

Department of Chemical Engineering, Indian Institute of Technology, Bombay, Powai, Mumbai 400076, India

Address correspondence to: Kartic C. Khilar, Department of Chemical Engineering, Indian Institute of Technology, Bombay, Powai, Mumbai 400076, India; E-mail: kartic@iitb.ac.in

Abstract: The water in oil microemulsion or reverse micelle has been used in the past two decades for the synthesis of many different types of nanoparticles. The nano meter sized aqueous cores of the reverse micelle provide an appropriate stabilized environment for the production of nanoparticles of fairly uniform size, through chemical reactions occurring in the core and it also acts as steric stabilizers to inhibit the aggregation of nanoparticles formed. The water in oil microemulsion has been used to synthesize different types of core nano particles (metals, and semiconductors) as well as core-shell/ composite nanoparticles. This article describes the preparation techniques, and the various techniques used to characterize these core and core-shell nanoparticles as well as insights in to the effects of various process parameters on the terminal particle size. A brief review of our modeling work based on stochastic population balance is also presented, which can be used to describe the formation of both core and core-shell nanoparticles. In addition, we have also presented a brief review of the work on the synthesis of anisotropic nanostructures like nanorods and nanowires by templating against surfactant micelles and reverse micelles. Some findings of our work, addressing the engineering issues, such as possibility of reusing surfactant and organic phases are also included in this article.

Key words: Microemulsion, micelle, nanoparticles, reverse micelles.

1. INTRODUCTION

Nanoparticles are small clusters of atoms having one or more dimension in the range of one to hundred nanometers, which causes their properties to be different from bulk materials. They form a bridge between atomic or molecular structures and bulk materials; therefore currently there is a great deal of scientific interest among them. Nanocomposite is a nano structure consisting of two or more materials in some forms and its dimension falls in the nano meter range in at least one direction.

Core-shell nanoparticles are a special type of nanocomposite material having a particular morphological structure of core particle surrounded by shell of different material.

Unlike bulk materials the nanoparticles physical properties depends not only on the size of the particles but also shape and morphology [1, 2], such as quantum confinement in semiconductor particles, surface plasmon resonance in some metal particles and supermagnetism in magnetic materials. For bulk materials larger than one micrometre, the percentage of atoms at the surface is minuscule as compared to the total number of atoms of the bulk material. As the size of the crystal is reduced, the number of atoms at the surface of the crystal as compared to the number of atoms in the crystal itself increases. The interesting and unexpected properties of nanoparticles are partly

due to the fact that surfaces of the material influence the so called bulk properties.

In case of metals and semiconductors, an important phenomenon known as size quantization effect is observed. The size of the nanoparticle is compared to the de Broglie wavelength of its charge carriers (electrons and holes). Due to spatial confinement of these charge carriers, the edge of the valence and conduction bands split into discrete quantized energy levels similar to those in atoms and molecules. The spacing of the energy levels as well as band gap of semiconductor increases with decreasing particle size. This increases in band gap which can be observed by the blue shift in the absorption spectrum.

In the case of metal nanoparticles, optical resonance known as surface plasmons are generated, when the oscillating electrons in the conduction band interact with an oscillating electric field. The resonating frequency depends on the size and shape of the nanoparticles. In metal nanoparticles, we can have interesting phenomena, like single electron tunneling and Coulomb blockade. These properties are utilized in designing some useful devices.

In magnetic nanoparticles, significant changes in magnetic properties occur. The coercivity (reverse field strength needed to reduce the magnetization to zero) depends on the size of the nanoparticles. The susceptibility (efficiency of the applied magnetic field

Sunil Kumar Bajpai/Murali Mohan Yallapu (Eds.)

to magnetize the materials) is also size dependent in the nanoscale range. In addition, the saturation magnetization has found to decrease with decrease in the size of particles. The size and shape of the magnetic nanoparticles influence the blocking temperature of the material (transition from superparamagnetic to ferromagnetic).

By the virtue of their novel and potentially useful properties, nanoparticles have created large amount of interest in the field of science, engineering and industrial applications. Nanostructured materials are used as key components in many areas [1-5] as electronics and optical devices, pharmaceuticals, paints, coatings, superconductors, drug delivery, biomedical applications and catalysis.

Several physical and chemical methods for the synthesis of nano particle have been developed in the last two decades [6, 7]. One of these techniques involves the use of reverse micelles or water in oil microemulsions of surfactants as "nano reactors" [8-14].

2. MICROEMULSIONS

A microemulsion can be defined as the thermodynamically stable, optically clear dispersion of two immiscible liquids (water and oil) consisting of nano-sized domain of one or both liquids, that are stabilized by an interfacial film of surfactants molecule.

Microemulsions are generally classified into two types – oil in water (O/W) and water in oil (W/O). In O/W type microemulsions, the bulk phase is water and these are called normal micelles. In W/O type of microemulsion, the bulk phase is oil and these are known as reverse micelles. The water in oil microemulsion (W/O), or reverse micelles consists of nano-sized water droplets which are dispersed in the continuous hydrocarbon phase and stabilized by surfactant molecules accumulated at the o/w interface. The main components of water in oil microemulsions are:

a)   Surfactant: AOT, CTAB, Triton X 100, SDS
b)   Co surfactant: Aliphatic alcohols with chain length of C6 to C8.
c)   Organic solvents: Alkanes or cycloalkanes with six to eight carbons.
d)   WaterR, the water to surfactant molar ratio, characterizes the water solubilized in the core, forming a 'waterpool'. ($R= [H_2O]/[S]$). The aggregates containing a small amount of water are called reverse micelles (R = 15), whereas the microemulsions consist of droplets with a large amount of water molecules (above R = 15)

Water in oil microemulsions can have different structures, like sphere, rods or disc shaped. Moulik and Paul have discussed the fundamental understanding of microemulsion properties, such as phase behavior, structures and dynamics in an excellent review article [15].

This thermodynamically stable dispersion, microemulsion can be considered as nano reactors, which can be used to carry out chemical reactions, and particularly synthesize different nanomaterials with new and special properties. A large variety of nanoparticles have been prepared by this method, since the invention of this technique by Boutonnet *et al.* [16] in 1982. They have demonstrated that the versatile methodology to prepare metallic nanoparticles (platinum, palladium and rhodium) by simple mixing of two water in oil microemulsions, one containing a salt or a complex of the metal and the other containing a reducing agent, such as sodium borohydride or hydrazine. A large number of review articles are now available on synthesis of nanoparticles in water in oil microemulsion [8-11, 17-19]. This technique has also been used for production of more complex core- shell and composite nanoparticles. (CdS- Ag_2 S [20], Co-Ag [21], Fe_3O_4- SiO_2 [22]).

2.1. Advantages of Using Microemulsions for Nanoparticles Synthesis

The W/O microemulsion can be used as nano reactors to produce nanoparticles by carrying out chemical reactions in their aqueous core and also prevents aggregation of particles by acting like cage. This technique does not require any special equipment and extreme conditions of temperature and pressure for processing. It is possible to control the size and morphology of particles formed. The nanoparticles can be stored in these W/O microemulsions for a long time without aggregation.

2.2. Nanoparticles Synthetic Routes in Water in Oil Microemulsions

Synthesis of nanoparticles can be carried out inside microemulsions by using the following routes:

Microemulsion plus trigger: In this method, microemulsion containing molecules of a reactant is subjected to external source of energy, such as UV rays, laser etc to produce nanoparticles. Kurihara *et al.* [23] have synthesized gold nanoparticles by subjecting a microemulsion containing gold chloride solutions to laser photolysis.

Two microemulsion method: Two identical microemulsions containing two reactants in their aqueous core are mixed. Collisions between droplets containing reactant A and droplets containing reactant

B leads to inter micelle mixing and reaction resulting in product molecules. This is the most popular method to synthesize nanoparticles by using water in oil microemulsions. Nanoparticles of Ag [24] and AgCl [25] and CdS and Ag $_2$S [20] have been synthesized by this method.

Single microemulsion method: In this method a microemulsion containing one of the reactant is used, while the other reactant is introduced into the continuous or oil phase. The reactant molecule in the continuous phase diffuses into the microemulsion core and reacts to produce the product molecules. TiO$_2$ nanoparticles have been prepared by this method by Ghodke *et al.* [26]. CdS nanoparticles have been prepared by bubbling H$_2$S through a microemulsion containing Cd (NO$_3$)$_2$[27].

2.3. Important Parameters Affecting Particle Formation in Water in Oil Microemulsions

Average Occupancy Number: Average number of reactant molecules present in the microemulsion droplets is known as average occupancy number. Experimental findings of Nagy [28] and Lianos and Thomas [29] have shown that small size particles are formed at high occupancy number and vice versa. This is because at high occupancy number, a large number of nuclei are formed, resulting in a large number of particles of smaller size. Large number of nuclei is formed as there will be higher number of microemulsion droplets having product molecules greater than the critical number required for nucleation, due to high initial occupancy and higher rate of nucleation. This result has been supported by simulation studies of Natarajan *et al.* [30]. The average occupancy number can be increased by using higher reactant concentration or lower surfactant concentration.

Nature of Surfactants: For a given composition (nature of oil and aqueous phase), the nature of surfactant molecule determines the exchange rate through the rigidity of the surfactant shell or interface. The rate of reaction of a fast reaction is controlled by the rate of coalescence, which depends upon the rigidity of the interface. Longer hydrophobic chains lead to more rigid interface, thus slowing intermicellar exchange. Lianos and Thomas [31] have studied the effect of various surfactants, like (Cetyl trimethyl ammonium bromide) CTAB, (sodium dioctyl sulfosuccinate) AOT, (sodium dodecyl sulfate) SDS on formation of CdS nanoparticles and they found that there is no effect on size on variation of surfactant indicating that rate of reaction is much slower than intermicellar exchange rate.

Some of the common surfactants used for nanoparticle synthesis are anionic surfactant AOT, cationic surfactant CTAB, or di-n- didodecyl - dimethyl ammonium bromide (DDAB) and nonionic surfactants Triton X100, polyoxyethylene (5) nonylphenyl ether (NP-5). A more detailed list of different surfactants investigated can be found in a review by Lopez-Quintala *et al.* [32]. Bagwe and Khilar [24] have reported that that by addition of small amount of non-ionic surfactant to the AOT/ heptane microemulsion for the synthesis of silver nanoparticles, the particle size decreases. It has been further shown by various studies [8] that the shape, size and stability of the nanoparticles depends upon the type of surfactant used.

Effect of addition of co-surfactant (intermediate chain length alcohols) on the size of nanoparticles has also been studied [24, 32, and 33]. Bagawe and Khilar [24] have studied the effect of variation of chain length and concentration of cosurfactant on formation of silver nanoparticles in cationic reverse microemulsion. They found a decrease in particle size when chain length was shortened. The addition of cosurfactant leads to higher fluidity of the intermicellar film thus leading to increase in rate of inter-micellar exchange.

Water to surfactant ratio (R): The water to surfactant molar ratio (R) can be increased in two ways. The first way is by keeping oil to surfactant ratio constant and increasing R by increasing water content. The second way is by changing surfactant concentration keeping the water content fixed. In case of AOT surfactant, the micellar size (r) is related to R by the relation [34]
r = 1.8R + 4.5, r is in A^0.

In the first case, the aggregation number of surfactant changes with R, leading to increase in micellar size and reduction in concentration of micelle. The ion occupancy also increases in this case. The particle size continuously increased with increase of water content. Above R= 10-15, the particle size does not change drastically but increase in polydispersity is observed. The increase in particle size with water content is attributed to the structure of water inside the waterpool. At low water content the number of water molecules per surfactant is too small to hydrate the counter ions and head polar groups. The water molecules can be considered bound, the micelle interface is said to be rigid lowering intermicellar exchange and growth rate. As R is raised, the film becomes more fluid. Bommarius *et al.* [35] have suggested that intermicellar exchange rate increases as R increases due to increase in flexibility of interface having larger radius of curvature, thus leading to larger growth rate. At R greater than 10-15, the extra water added is just added to bulk waterpool, leading to dilution of reactants decreasing reaction rate, and effect of increase in intermicellar exchange rate is negated. In some cases even decrease in particle size is observed at higher R values.

In the second case, the increase in surfactant concentration causes decrease in R value and size of droplet but increases the number of droplets. Qi et al. [36] have found that increase in surfactant concentration keeping water content fixed causes decrease in size and polydispersity of $BaCO_3$ particles.

Effect of Oil Phase: Different groups have investigated the effect of oil phase on growth of nano particles showing that different oil phases affect intermicellar exchange, particle size and polydispersity. Bagwe and Khilar [25] have shown that in case of AgCl nanoparticles formed in AOT, the longer is the chain length of the oil, more difficult is the penetration into the surfactant tail and hence there is a tendency to align itself parallel to the surfactant tail. Such alignment increases the intermicellar exchange rate. As a result, particle sizes decrease.

Effect of cations and additives: It has been observed that size of the particle is weakly affected by changing the cation of the salt. Bagwe and Khilar [25] have shown that smaller number of large sized AgCl particles are formed in the presence of higher charged (small size) cations at the interfacial region, which in turn inhibits materials exchange and adversely affects the particles formations. It has been shown [24] that effects of additives on shell rigidity can alter the micellar size. For example benzoyl alcohol and toluene have been found to decrease the micellar size and hence increase the number density of micelles as a result, ion occupancy number decreases, which in turn, can affect the particle size.

Intermicellar Exchange: The intermicellar exchange rate is found to affect both the terminal particle size and the rate of particle formation [32]. In the experimental measurements of Bagwe and Khilar [24-25], in majority cases, conditions of higher intermicellar exchange rate produced smaller size particles but not always. Slower exchange conditions in the majority cases produced larger particle size.

It has been observed that ionic species present in reverse micelles display chemical reactivity altogether differ from bulk aqueous phase. These effects are highly specific to chemical species type of surfactant and reaction itself. For example, in the synthesis of Ni_2B, Nagy et al. [37] observed that Ni (II) ions are linked to two hexanol molecules at the surfactant / water interface in CTAB/ hexanol /water by conducting C^{13}- NMR experiments, while Co (II) ions form octahedral complex at the interface containing one hexanol molecule. Fe(II) is strongly hydrated and is located away from the interface. During the particle preparation it was observed that Ni_2B and Co_2B particle size went through a minimum, while Fe_2B steadily increased. Nagy et al. [37] also synthesized Co_2B in Triton X 100 /decanol/ water system, where it

was found that size of the particles were independent of metal ion concentration and microemulsion composition. It has been suggested that in non ionic microemulsion Co (II) ions are solubilized in aqueous as well as organic phase.

3. PREPARATION OF CORE NANO-PARTICLES

A variety of nanoparticles have been prepared in water in oil microemulsion. A large number of review articles are available in the literature on nanoparticle synthesis in reverse micellar medium [8-11,17-19]. Table **1** summarizes some of the nanoparticles synthesized by using water in oil microemulsion. In this section, we briefly summarize the preparation technique and results of core nanoparticles prepared by our group.

Table 1.   Different Types of Nanoparticles Formed in Reverse Micellar Media

Metals	Pt [16], Cu [38, 39] Co [40], Ag [24], Ni. [41],
Semiconductors	ZnS [42], CdS [43], PbS[44], CdTe [45]
Metal oxides	TiO_2 [26, 11], SiO_2 [46], Fe_2O_3 [47],
Other inorganic Compounds	$CaCO_3$ [48], AgCl [25]

3.1. Preparation of Nanoparticles by Two Microemulsion Method

Two microemulsions were prepared separately; by dissolving aqueous solution of the first reactant, to a fixed volume of reverse micellar solution of AOT in an organic solvent, like heptane, and dissolving the aqueous solution of the second reactant to an equal volume of AOT reverse micellar solution. The two microemulsions were then mixed to produce a third microemulsion containing the nanoparticles. In all the cases, transparent microemulsions were obtained. The water to surfactant molar ratio R was maintained constant in each set of experiments.

The preparation of nanoparticles by using two microemulsions method involves the following steps. Fig. **(1)**

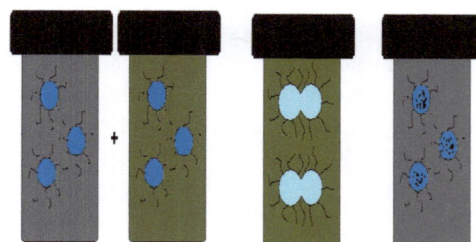

Fig. (1). Synthesis of nanoparticles by two microemulsion method.

1. Equal volumes of two microemulsions containing two different reactants in their aqueous core are mixed.
2. The Brownian motion of the reverse micelles leads to collision.
3. The surfactant layers open up and coalescence occurs in forming transient dimers. (Fusion).
4. Mixing of reactants during fusion.
5. Reaction between reactants, giving rise to products.
6. Decoalescence to return as reverse micelles (Fission)

When two microemulsions are mixed, assuming instantaneous reaction, the rate of formation of product molecules depends upon the intermicellar exchange rate coefficient (k_{ex}) [24, 25]. Furthermore, the rate of accumulation of product molecules in a reverse micelle determines the number of micelles that may have number of molecules beyond the critical nucleation number to nucleate a particle. The terminal particle size distribution and number density of nanoparticles therefore, depend on the intermicellar exchange rate. The effects of intermicellar exchange rate on the terminal size of some metal and metal salt nanoparticles have been studied [24, 25].

The values of k_{ex} for n-heptane and n-decane differ by an order of magnitude and are equal to 10^7 and 10^8 M^{-1} s^{-1}, respectively [24]. The presence of additives, such as benzyl alcohol and toluene also changes the exchange rate coefficient. Such increase or decrease in the exchange rate coefficient is primarily due to the change in curvature and rigidity of surfactant shell induced by the interaction of surfactant molecules with the molecules of the organic phase as well as with molecules of additives if present.

The effect of organic solvents and additives on terminal size of nanoparticles of silver chloride and silver were studied. Here the size refers to the diameter (dp) of the nanoparticles. It was found that the size of silver chloride nanoparticles increased as the organic phase was changed from decane to hepatane to cyclohexane and thereby changing k_{ex} from 10^8 to 10^7 to 10^6 $M^{-1}s^{-1}$.

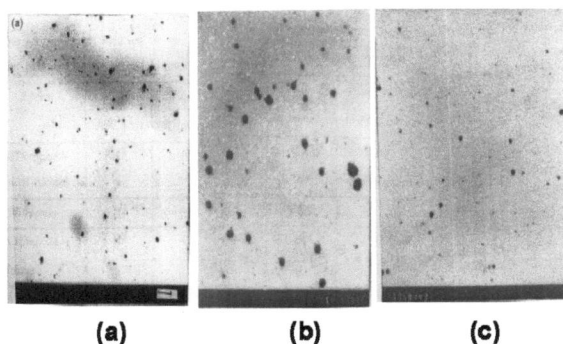

(a)   (b)   (c)

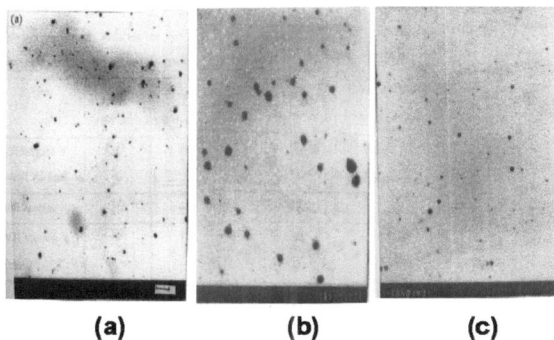

(a)   (b)   (c)   Fig. (2). TEM of silver nanoparticles showing the effect of organic additives: (a) AOT/heptane + benzyl alcohol, (b) AOT/heptane and (c) AOT/heptane + toluene.

Figs. (**2a-c**) show the TEM micrographs of silver nanoparticles in the presence of additives. We observe from Figs. (**2a**) to (**2c**) that the size of silver nanoparticles also changes with the presence of additives in heptane due to the change in the values of exchange rate coefficient. Water to surfactant molar ratio R is known to affect the particle size [24, 25]. The average size of silver chloride [25] and cadmium sulfide [20] nanoparticles are presented in Table **2**.

We observe from Table **2** that, while the water to surfactant molar ratio, R has a significant effect on the size of silver nanoparticles formed, its effect on the size of cadmium sulfide nanoparticles formed is insignificant. As R changes, both the surfactant layer rigidity and ion occupancy alter, in the most cases in an opposite manner, resulting in a complex effect. Some studies have been conducted, where the rigidity of the surfactant shell has been modified by some means, such as changing the salt from NaCl to $CaCl_2$ in case of formation of silver chloride nanoparticles and adding a minute amount of other surfactants to AOT micelles in case of formation of silver nanoparticles [24, 25]. The number density is found to decrease by a factor five, arising out of formation of relatively rigid surfactant shell [12]. In the other set of experiments, a small amount of surfactant (less than one tenth amount of AOT) such as SDS, DTAB or NP-5 was added. It has been found that that the average size of nanoparticles formed is around 20 nm excepting in case of NP-5, where it is found to be 7.0 nm. [24]. The oxyethylene units of NP-5 are believed to act as cosurfactant bringing about the fluidity of the surfactant shell, which, in turn increases the inter-micellar exchange rate.

Table 2. Effect of the Water to Surfactant Molar Ratio on the Average Particle Size

R	(AgCl) dp (nm)	(CdS) dp (nm)
5	5.7	3.0
0	10.1	3.2
15	5.7	3.4

3.2. Preparation of Nanoparticles Using Single Microemulsion Method

Titanium dioxide nanoparticles were prepared by this method [26]. Solution I was prepared by adding Triton X 100, cyclohexane and n-hexanol in a proportionate amount. Solution II was prepared by adding the required volume of $TiCl_4$ to a calculated volume of the cyclohexane under constant stirring. This solution was a clear yellow solution. Solution I was added to Solution II to obtain a clear transparent Solution III. Aqueous solution of ammonia was then added drop wise to solution III. This leads to swelling of reverse micelles as the ammonia is not soluble in the organic phase and hence enters the reverse micelles almost immediately. The precipitate formed instantaneously due to reaction between ammonia and $TiCl_4$. The solution was then centrifuged for 15 minutes at 10,000 rpm using a REMI C-24 Centrifuge at room temperature. The particles were then washed with a 50:50 v/v mixture of methanol and chloroform and centrifuged again for 15 minutes at 10,000 rpm to remove the surfactant. The solid was then removed and dried for 4 hours in an oven at 80-100 °C.

Single microemulsion techniques while widely used for systems, where one of the reactants is a gas, have not similarly been exploited for systems involving liquid phase reactants. We have demonstrated the use of a single microemulsion method [26] to produce highly monodisperse TiO_2 nanoparticles using the organic phase as one of the reactant phases. Single microemulsion methods are important because by using the continuous phase for dispersing one of the reactants, use of expensive surfactant can be minimized and also, they offer reuse capabilities.

Fig. (3) shows typical TEM images [26] of the TiO_2 nanoparticles obtained using single microemulsion method, at a magnification of 66,000 and 1,15,000, respectively. The TEM images show a large number of very small single nanoparticles (2-5 nm) as it can be seen in Fig 3a. We also observe the presence of large agglomerates in the size range of 100 - 200 nm. We gather from these images that the large agglomerates are clusters of smaller particles characterized above. TEM images of Nano TiO_2 also reveal the presence of a second type of nanoparticle. We also observe that few relatively large (10 – 30 nm), irregularly shaped nanoparticles are also present. Based on the observation of two types of particles in the TEM images it is hypothesized that these two types of particles could possibly originate in two different phases – the organic and aqueous phases of the reaction mixture.

Using XRD, we demonstrated that reaction in the continuous phase occurs to a significant extent during the process of swelling of microemulsion giving rise to pseudobrookite nanoparticles, whereas reaction inside the aqueous cores of the reverse micelles leads to the synthesis of rutile TiO_2. The presence of two types of nanoparticles indicates that there are two dominant mechanisms for nanoparticle synthesis. We propose that some of the ammonia may react with the $TiCl_4$ in the continuous phase during the process of micelle swelling, giving rise to organic phase TiO_2 nanoparticles. (pseudobrookite phase). The ammonia present in the swollen reverse micelles then reacts with $TiCl_4$ to produce small nanoparticles corresponding to aqueous phase TiO_2 (rutile phase).

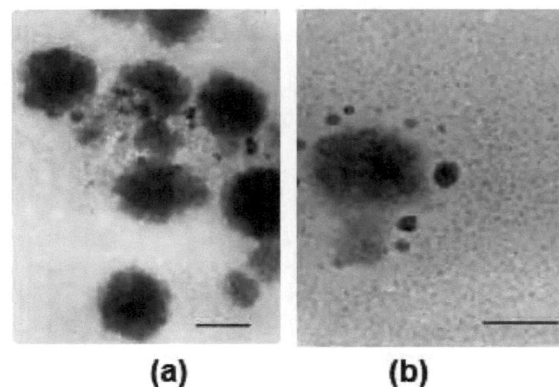

(a)   **(b)**

Fig. (3). TEM images of pseudobrookite titania. Scale bar: (a) 200 nm and (b) 100 nm.

4. PREPARATION OF CORE-SHELL NANOPARTICLES

One of the main advantages of the microemulsion method to prepare nanomaterials, over other preparation method, is the ability to control the formation of different kind of core-shell nanoparticles [18, 20]. The core-shell nanoparticles were prepared by Hota and Khilar [20] in microemulsions of AOT in heptane. $CdS-Ag_2S$ core-shell nanoparticles were synthesized by using two mixing methods:

1) Post Core Method: A micro emulsion of $AgNO_3$ solution was added to the microemulsion containing CdS nanoparticles, and an excess amount of $(NH_4)_2S$ which act as cores.

2) Partial Micro Emulsion Method: $AgNO_3$ solution was added drop wise directly to the microemulsion containing core CdS nanoparticles with excess $(NH_4)_2S$. The particles thus produced have been characterized by TEM, EDAX, Photoluminescence, XPS and UV-Visible spectrophotometer. Fig. (4) presents the transmission electron micrographs for Ag_2S coated CdS nanoparticles at different values of R. We observed from the micrographs that the core-shell morphology has not been captured due to poor contrast arising out of the fact that both core and shell materials are similar in nature, particularly in electron density. The indirect evidence, such as XPS and

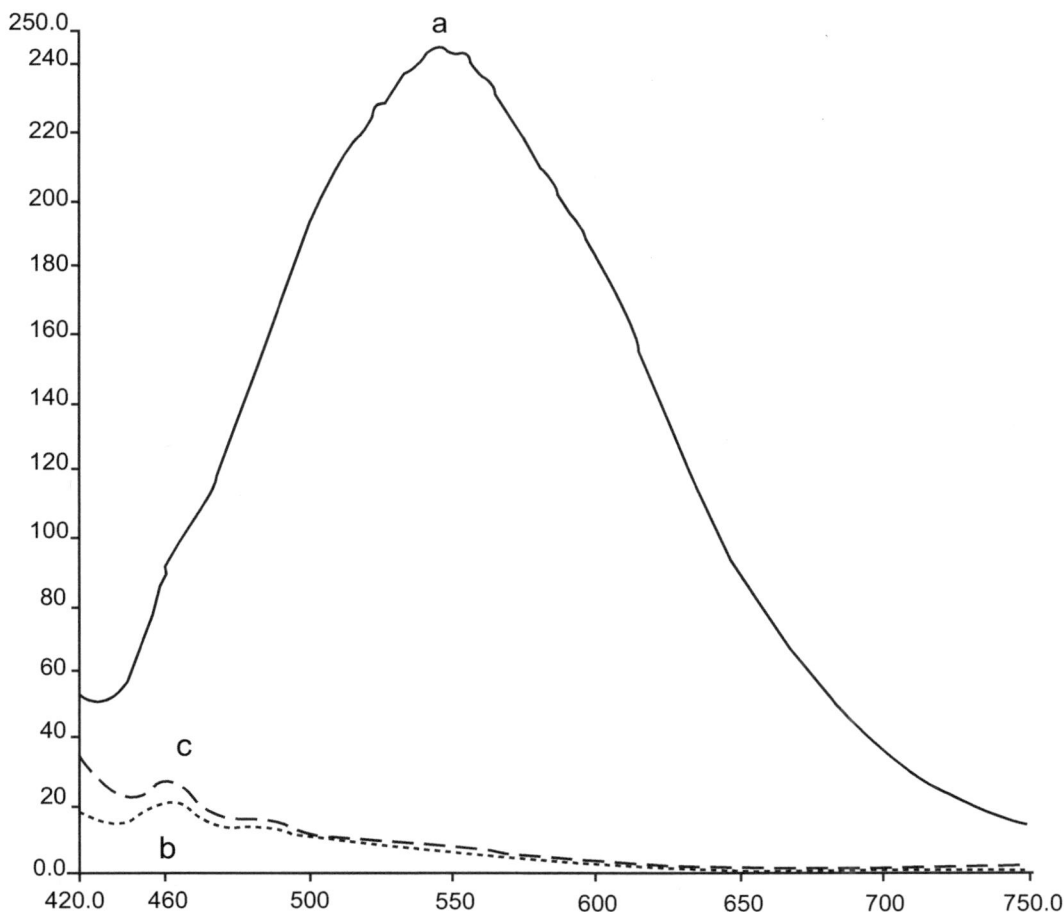

Fig. (5). Fluorescence spectra of (**a**) CdS; (**b**) Ag$_2$S; and (**c**) CdS@Ag$_2$S nanoparticles prepared by w/o microemulsions at room temperature.

photoluminescence measurements strongly indicates core-shell morphology [49].

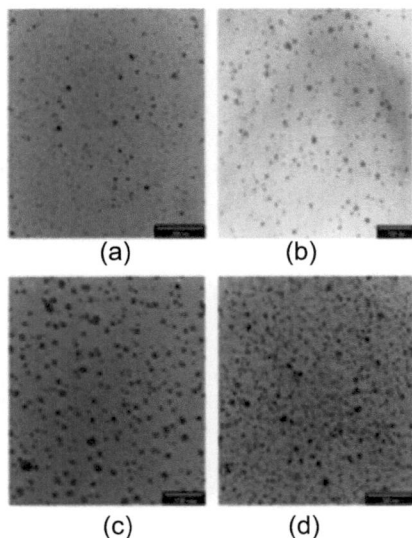

Fig. (4). TEM micrograph of CdS @ Ag$_2$S nanoparticles at different R. (**a**) R=5, (**b**) R=10, (**c**) R=15, and (**d**) R=20.

The atomic weight ratio of Ag to Cd as estimated from the XPS spectra of core- shell particles clearly suggests that the CdS particles are covered with a layer of Ag$_2$S material. Support for the core shell formation of CdS @ Ag$_2$S is also given by fluorescence spectra Fig. (5). The fluorescence spectra of CdS is found to be suppressed in CdS @ Ag$_2$S particles due to formation of shell of Ag$_2$S around CdS core nanoparticles The average size of the formation of coated nanoparticles as measured is approximately 6.0 nm which is higher than the average size of 3.0 nm for the core nanoparticles of CdS. (Table **3**) [20].

Table 3. **Particle size distributions of CdS @ Ag$_2$S nanoparticles at different values of R**

R	dp (CdS@Ag2S) nm	dp (CdS) nm	Shell Thickness nm
5	4.8	3.0	0.90
10	6.6	3.2	1.70
15	7.7	3.4	2.15
20	8.1	-	

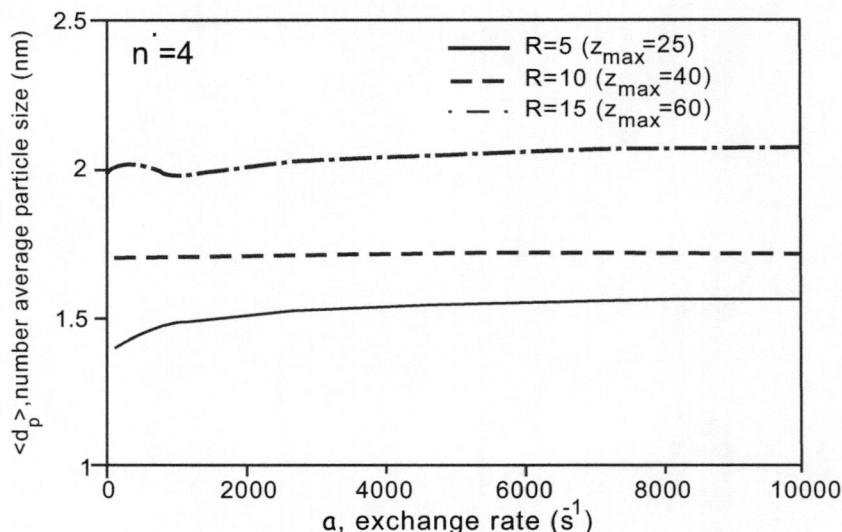

Fig. (6). Effect of exchange rate on the number average particle size for different values of water-to-surfactant molar ratio.

Silica coated Ag and Ag$_2$S nano particles were also prepared by partial microemulsion method [50]. It has been concluded from all these studies [20, 49, 50] that partial micremulsion method is a better technique for producing core- shell nanoparticles.

5. MODELING OF NANOPARTICLE FORMATION USING STOCHASTIC POPULATION BALANCE APPROACH

5.1. Core Nanoparticles

A stochastic population balance model for the formation of nanoparticles, involving mixing of two micellized aqueous solutions each containing a reactant which has been developed to study the effects of various parameters that influence the formation process of nanoparticles. The details of the model, including the stochastic population balance equations, nucleation rate, population balance equations, Poisson distribution, numerical methods and the parameter values are given elsewhere [51]. The model predictions are found to qualitatively agree with experimental measurements [14, 51]. The parameter sensitivity exercise has been conducted in details using model simulations and only the one related to the exchange rate is presented.

The effects of intermicellar exchange rate on the particle size for different water to surfactant molar ratio (different ion occupancy) as shown in Fig. (6). implies that, in general, the effect is weak. This is expected, as at high exchange rate, the phenomena of nucleation and growth is not controlled by the exchange process. Interestingly, at low exchange rate, the effects of exchange rate are dependent on the ion

occupancy as determined by R in this simulation. At high ion occupancy (high value of R) the particle size decreases. At low ion occupancy (low value of R), in fact, the particle size increases. At moderate values, the particle size is negligibly affected by the exchange rate. These observations can be explained by considering the mediation of exchange rate in mixing of reactants to produce product molecules and assuming that the product molecules need to accumulate to nucleate a particle.

Good amount of quality modeling work based on a different approach, Monte Carlo (MC) simulation has been conducted by Bandyopadhyaya *et al.* [52], Ethayaraja *et al.* [53], and Shukla and Mehra [54]. The predictions of these models agree reasonably with the experimental measurements. Essentially, with the same physics, both approaches differ in the requirement of computational time. Furthermore, MC approach has a synergetic link with random and stochastic processes and, thereby, may be a preferred one if computational time is not prohibitive.

5.2 Core-Shell Nanoparticles

Core and shell nanoparticle formation in reverse microemulsions has been formulated by taking the post-core method of synthesis, using population balance equations. The formation of core nano particle was simulated by a modified model of Kumar *et al.* [51]. Formation of core and shell nano particles was followed, thereafter, by considering the probabilities of micelles of core nanoparticle formation, which were most significant. The study [55] provides an insight into the effect of the concentration of reactants on the shell coating and shows that there exist nanoparticles which are completely coated, nanoparticles which are

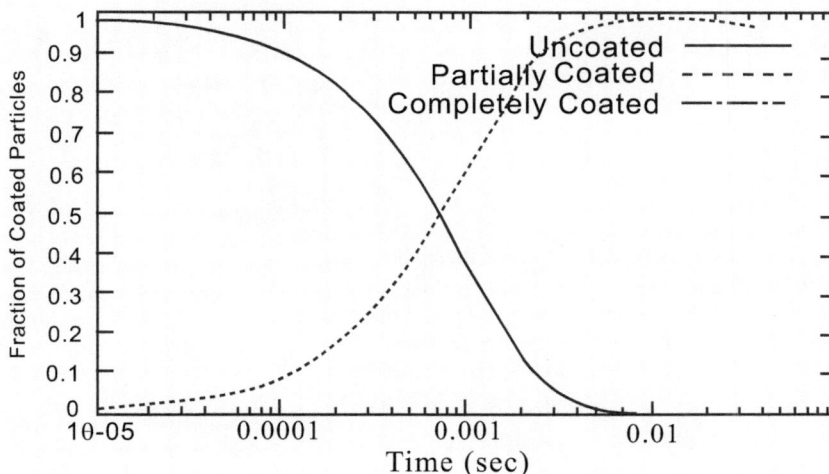

Fig. (7). Percentage variation of coating for Cs (**A**) = 0. 2 M , Cs (**B**) = 0.1 M, Cs (**C**) = 0.1 M.

partially coated and nanoparticles with no coating at all. It is shown that there is a large fraction of nanoparticles which are partially coated.

In the post-core method, coated nano particles of core material Z and shell material Z1 can be prepared by using the chemical reactions A + B $\longrightarrow$ Z and A + C $\longrightarrow$ Z1. Firstly, a microemulsion containing reactant B molecules and another containing excess of A reactant molecules are mixed to produce a microemulsion of core nanoparticles of Z containing additional A molecules. To this, a microemulsion containing C molecules is added to prepare core-shell nanoparticles. Therefore, this method involves three microemulsions. Simulations were carried out with concentration of reactants taken as 0.2 M of A, 0.1 M of B and 0.1 M of C. These were followed by a number of simulations by keeping a concentration of B constant and increasing that of A. These simulations were conducted to find out the increase in the percentage of completely coated particles with the increased presence of shell material. The model calculations show that there exist not only completely coated particles, but also a large fraction of

partially coated ones. With an increase in reactant concentration, there was an increase in the percentage of completely coated nanoparticles. The details of this work are given elsewhere [55]. Here, we have illustrated only the variation of percentage coating with time.

Fig. (7) shows that the probability of particles with no coating decreases with time, whereas there was an increase in the number of partially coated particles. Moreover, the completely coated nanoparticles were very meager with less than 2% of the particles being completely coated.

6. REUSE OF SURFACTANT/OIL PHASE

As pointed out earlier, there are several advantages of using the microemulsion for nanoparticle synthesis; but one constraining factor is that the amount of nano particles formed is very small compared to the amount of surfactant / oil phase used. The cost of surfactant /oil phase constitutes almost 90% of the total cost of

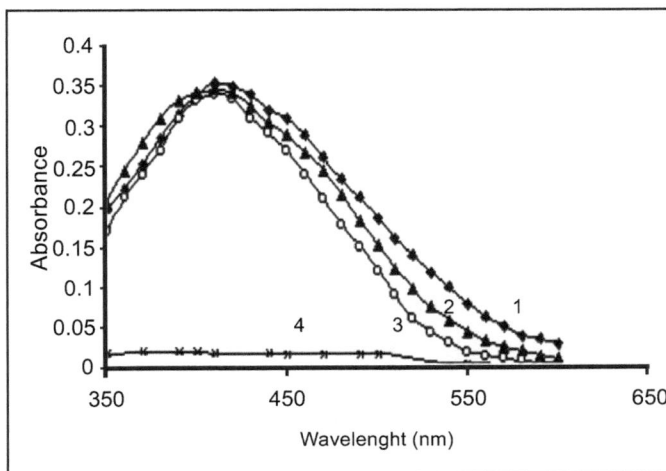

Fig. (8). UV–visible spectra of silver nano particles. 1 (Initial AOT/heptane /water system); 2 (1st Reuse); 3 (2nd Reuse); 4 (filtrate).

nano particles synthesized by this method. This calls for a detailed look into the possibility reusing the surfactant/oil phase to produce different batches of nanoparticles [56]. A method has been developed to separate the nano particles from surfactant/ oil phase , and reuse the surfactant oil phase for synthesis of different batches of nanoparticles, charecterize them and compare the charecteristics of nanoparticles produced by reused surfactant / oil phase. Silver nanoparticles were synthesized in water-in-oil microemulsion of AOT in heptane. These particles were characterized by UV-visible spectroscopy Fig. (**8**) (Ag nano particles give a peak at 410 nm.) and TEM Fig. (**9**).

The average particle size was noticed as 18.2 nm by TEM. The nanoparticles were separated by ultracentrifugation [56]. The centrifuge was used at a *g* value of 50,000 which corresponds to 31,250 rpm. The centrifugation was carried out for a total time of 1 h. The operating cost of separation by ultracentrifuge (electricity cost) was less than 1% of the total cost of nanoparticle synthesis by this method. The filtrate was reused twice to produce second and third batches of silver nanoparticles. The UV-visible spectra of second and third batch of silver nano particles also showed peaks at 410 nm Fig. (**8**). The TEM micrographs of silver nanoparticles synthesized after first Fig. (**9b**) and second reuse Fig. (**9c**) show that the shape of nanoparticles formed by the reused surfactant/oil phase remains virtually same as fresh surfactant/organic phase. The average size of the particles as determined by the software Image pro plus after first and second reuses was 18.7 and 18.2 nm, respectively.

We can conclude from these TEM micrographs and UV-visible spectra that the geometric characteristics and the chemical content of the nanoparticles formed by reused surfactant/organic phase remain virtually the same as those in case of fresh surfactant/organic phase. Moreover, the nanoparticles remain stable in the microemulsions made of fresh as well as reused surfactant for more than a month. Therefore, the reuse of surfactant and organic phase is a good possibility for large scale synthesis of nanoparticles by this method.

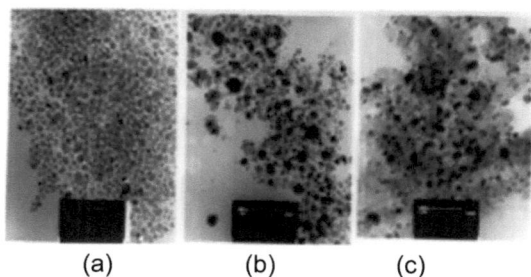

(a) (b) (c)

Fig. (9). TEM scans for silver nano particles in (**a**) Initial microemulsion; (**b**) First Reuse; (**c**) Second Reuse. The scale bar represents 100 nm.

7. ANISOTROPIC NANOSTRUCTURES BY TEMPLATING AGAINST SURFACTANT MICELLES AND REVERSE MICELLES

Currently there is great interest in nanorods and nanowires [4, 57-64]. Different workers have shown that anisotropic materials show improved electronic, magnetic and mechanical properties as compared to nanospheres. The optical properties of metallic nanorods have been shown to depend upon the shape.

Different methods have been used for the synthesis of nanowires and nanorods [4], such as the use of capping agents, use of hard templates (alumina membrane, existing nanostructures, or features on solid surfaces), solvothermal methods, vapor-liquid-solid method and so on.

The non-spherical micelles and reverse micelles of surfactants like rod shaped and disc shaped ones can be used as soft templates for synthesis of anisotropic structures like nanorods and nanowires. The aspect ratio of the resulting nanorod is controlled by shape and size of microemulsion template, concentration of salt, precursors and surfactants.

Murphy and Jana [57] have developed a seed mediated growth approach to synthesize metallic nanorods and nanowires by using rod like micelles of cationic surfactant cetyl trimethylammoniumbromide (CTAB) as template. They have first synthesized 3-5 nm gold and silver nanospheres by reduction of metallic salt with $NaBH_4$. These seeds were then added to a solution containing more metal salt, a weak reducing agent like ascorbic acid and rod like CTAB micellar templates. The seeds serve as nucleation sites for growth of nanorods using the CTAB micellar nanorods as templates. Nanowires were produced by fine tuning of solution conditions. The aspect ratio was controlled by ratio of metal seeds to metal salts.

Kwan *et al.* [58] have synthesized large aspect ratio $BaWO4$ nanorods by using reverse micelle templating method. Barium bis (2 –ethylhexyl) sulfosuccinate [Ba $(AOT)_2$] reverse micelles were added to NaAOT microemulsion droplets containing sodium tungstate ($Na2WO_4$) at [H_2O]/[NaAOT] molar ratio =10, giving rise to a white precipitate of $BaWO4$ consisting of nanorod arrays of 1500 nm length and 9.5 nm diameter.

An electrochemical method has been reported for the synthesis of gold nanorods stabilized by a mixed cationic surfactant system [59]. Huang *et al.* [60] have synthesized Silver nanowires by electrodeposition from liquid crystalline phases. First a reverse hexagonal liquid crystalline phase was prepared consisting of 1.4 M AOT, an oil phase of p-xylene and

0.1 M AgNO$_3$ as aqueous phase. The electrodeposition was conducted by using a potentiostat in a two electrode configuration with two narrowly spaced electrodes and the liquid crystalline phase as electrolyte. The silver nanowires of tens of micrometer length with uniform wire diameter of 20-30 nm deposited on the cathode. They concluded that the high electric field during electro deposition enhances alignment of liquid crystals.

Esumi *et al.* [61] have prepared anisotropic rod like gold nanoparticles by UV irradiation in rod like micellar solution of cationic surfactant hexadecyl trimethyl ammonium chloride (HTAC) as template. It was found from static light scattering that rod like micelles of HTAC were formed at concentration exceeding 25 wt % of HTAC in aqueous solution. In their study they used a 30 wt% aqueous solution of HTAC containing various concentration of HAuCl4 (0.5 – 40 mmol dm-3.) They have found that at the concentrations of HAuCl$_4$ 5 mmol dm-3 and above rod like gold nanoparticles were observed on UV irradiation. The length of the rods was found to increase with increasing UV irradiation time.

Kameo *et al.* [62] have prepared fiber like gold nanoparticles by UV irradiation of HAuCl4 aqueous solutions in the presence of cationic surfactant with various chain lengths. They concluded that formation of fiber like gold nanoparticles using cationic surfactants as templates depends on the concentration of alkyl trimethyl ammonium chloride (Cn TAC) and length of the fiber like gold nanoparticles increases with increase in alkyl chain length of Cn TAC. Li *et al.* [63] have synthesized three different Barium chromate nanostructure–linear chains, rectangular superlattices and long filaments as function of reactant molar ratios by fusing AOT microemulsion droplets containing fixed concentration of barium and chromate ions, respectively.

Li and co workers [64] have synthesized single crystalline tungsten (W) nanowires by templating WO$_4^{2-}$ ions against the lamellar phase of CTAB, followed by pyrolysis in vacuum. Metal and semiconductor nanowires have been synthesized in large quantities using surfactant micelles / reverse micelles as templates. One disadvantage of this method is that the removal of micellar phase is often difficult and time consuming.

8. CONCLUSIONS

A large variety of nanostructures have been synthesized by using water in oil microemulsions. This is a soft technique and does not require any extreme conditions of temperature and pressure. Another important advantage of this method over other methods is that one can easily control the formation of different types of core-shell nanostructures.

The non spherical micelles and reverse micelles of surfactants like rod shaped and disc shaped ones can be used as soft templates for the synthesis of anisotropic structures like nanorods and nanowires.

Particle growth has been shown to be dependent on intermicellar exchange rate. The important factors which control the nanoparticle size are nature of surfactant/cosurfactant, oil phase, and water to surfactant molar ratio (R), reactant concentration and presence of additives.

One disadvantage of this method is that, compared to the amount of surfactant/ oil phase used, the yield of nanoparticles is low. The cost of surfactant/ oil phase used constitutes 90 percent of the cost of nanoparticles synthesized by this method. In an attempt to overcome this shortcoming, we have made some studies on the reuse of surfactant/oil phase to synthesize silver nanoparticles. We conclude that the reuse of surfactant oil phase is a good possibility but this method needs to be tested for different types of nanoparticles using different reverse micellar media.

9. ACKNOWLEDGEMENTS

DS gratefully acknowledges the financial support of DST project 06 DS 024 and Centre for Research in Nanotechnology and Science, I.I.T., Bombay for providing infrastructure facility. KCK gratefully acknowledges the research contribution of his students Rahul Bagwe, Garuda Hota, Upendra Natarajan, Kalyan Houdique, A Rameshkumar, B Vishwanath, S. Jain, Sonia Tikku, H.B.Ghodke, Rahul Keswani, Piyush Gupta, Anand Gautam and Swati Rao.

10. REFERENCES

[1] Nalwa HS Eds. Handbook of Nanostructured Materials and Nanotechnology. NewYork: Academic press, 2000.

[2] Edelstien AS, Cammarata RC Eds. Nanomaterials: Synthesis, Properties and Applications. Philadelphia: Institute of Physics, 1996.

[3] Brigger I, Dubernet C, Couvreur P. 2002, Nanoparticles in cancer therapy and diagnosis. Drugs Delivery Reviews 2002; 54: 631-651.

[4] Xia Y,Yang P, Sun Y, Wu Y, Mayers B, Gates B ,Yin Y, Kim Y, Yan H. One Dimensional Nanostructure: Synthesis, characterization, and applications. Adv. Mater. 2003; 15: 353-389.

[5] Ozin GA, Nanochemistry: Synthesis in diminishing dimensions. Adv. Mater. 1992; 4: 612-649.

[6] Gleiter H, Nanostructured Materials: Basic concepts and Microstructure. Acta Mater.2000; 48: 1-9.

[7] Burda C, Chen X, Narayanan R, El Sayed M A , Chemistry and properties of nanocrystals of different shapes. Chem. Reviews.2005; 105: 1025-1102.

[8] Eastoe J, Hollamby M J, Hudson L, Recent advances in nanoparticle synthesis with reverse micelles. Adv. in Colloid and Interface Sci. 2006; 128-130:5-15.

[9]   Pileni M P., Reverse micelles as microreactors. J Phys Chem. 1993; 97: .6961- 6973.

[10]   Eastoe, J, Warne B. Nanoparticles and polymer Synthesis in microemulsions. Current Opinion in Colloid Interface Sci.1996; 1: 800-805.

[11]   Holmberg K, Surfactant templated nanomaterials synthesis, Journal of Colloid and Interface Sci. 2004; 274: 355-364.

[12]   Petit C., Lixon P, Pileni M P. Synthesis of nanosize silver sulfide semiconductor particles in reverse micelles, J Phys.Chem 1993; 97:12974-12983.

[13]   Fendler J H. Atomic and molecular clusters in membrane mimetic chemistry. Chem. Rev. 1987; 87:877-899.

[14]   Pinna N, Willinger M, Weiss K, Urban J, Schlögl R Local Structure of Nanoscopic Materials: V_2O_5 Nanorods and Nanowires .Nano Lett, 2003.; 3: 1131-1134 .

[15]   Moulik S P, Paul B K. Structure dynamics and transport properties of microemulsions. Adv .Colloid Interface Sci. 1998; 78: 99-195.

[16]   Bouttonnet M, Kipling J, Stanius P. The preparation of monodisperse colloidal metal particles from microemulsions. Colloids Surf. 1982:5: 209-225.

[17]   Pileni MP. The role of soft colloidal templates in controlling the size and shape of inorganic nanocrystals. Nature Mater. 2003; 2: 145-150.

[18]   Lopez-Quintela MA. Synthesis of nanomaterials in microemulsions: formation mechanism and growth control. Current opinion in colloid and Interface sci 2003; 8:137-144.

[19]   Bouttonnet M, Logdberg S, Svensson E E. Recent developments in the applications of nanoparticles prepared from water in oil microemulsions. Current Opinion in Colloid and Interface sci.2008; 13:270-286.

[20]   Hota G., Jain S., Khilar KC .Synthesis of CdS- Ag_2S Core-shell/Composite Nanoparticles using AOT/n-heptane /Water Microemulsions, Colloid Surf, Physiochemical Aspect 2004; 232: 119-127.

[21]   Rivas J, Sanchez RD, Fondado A, etal. Structural and magnetic characterization of Co particles coated with Ag. J Appl. Phy 1994; 76: 7564-6.

[22]   Santra S, Tapec R, Theodoropoulou N, Dobson J, Hebard A, Tan W. Synthesis and characterization of silica-coated iron oxide nanoparticles in microemulsions. Langmuir 2001;17: 2900 –29006.

[23]   Kurihara K, Kizling J, Stenius P, Fendler JH. Laser and pulse radiolytically induced colloidal gold formation in W/O microemulsion. J Am Chem Soc. 1983; 105:2574-2579.

[24]   Bagwe R P, Khilar KC. Effect of intermicellar exchange rate on the formation of silver nano particles in reverse microemulsions of AOT.Langmuir2000; 16: 905-910.

[25]   Bagwe R P, Khilar K C Effect of intermicellar exchange rate and cations on the size of silver chloride nanoparticles formed in reverse micelles of AOT. Langmuir 1997; 13: 6432-6436.

[26]   Ghodke H B, Keswani R K., Sarkar D, Khilar K C Synthesis of titanium dioxide nanoparticles using a single microemulsion method (communicated).

27]   Suzuki K, Harada M, Shioi A. Growth Mechanism of CdS ultra-fine particles in water in oil microemulsions .J Chem.Engg.Jpn 1996; 29:264.

[28]   Nagy JB. Multinuclear NMR characterization of microemulsions: preparation of monodisperse colloidal metal boride particles. Colloids and Surfaces 1989; 35: 201.

[29]   Lianos P, Thomas J K. Small CdS particles in reverse micelles .J Colloid Interface Sci. 1987; 117:505-511.

[30]   Natarajan U, Handique K, Mehra A, Bellare J R, Khilar K C Ultrafine metal particle formation in reverse micellar system: Effect of intermicellar exchange on formation of particles. Langmuir 1997; 12, 2670 - 2678.

[31]   Lianos P, Thomas J K. Colloidal metal particles of extremely small dimensions roduced in reverse micelles. Mat.Sci. Forum 1988; 25-26: 369-376.

[32]   Lopez-Quintela MA, Tojo C, Blanco MC, Garcia Rio L, Leis JR. Microemulsion dynamics and reactions in

microemulsions. Curr. Opinion Colloid Interface Sci 2004; 9: 264.-270.

[33]   Uskovik V, Drofenik M. Synthesis of materials within reverse micelles. Surf Rev Lett 2005; 12:239-245.

[34]   Monnoyer Ph, Fonseka A, Nagy JB Preparation of colloidal AgBr particles from microemulsions. Colloids Surf.: Physico-chemical and Engineering Aspects 1995; 100 :233-243.

[35]   Bommarius A.S, Holzwarth JF, Wang DIC, Hatton TA Coalescence and solubilizate exchange in a cationic four-component reversed micellar system J Phys.Chem. 1990; 94: 7232-7239.

[36]   Qi L, Ma J, Cheng H, Zhao Z. Reverse micelle based formation of $BaCO_3$ nanowires. J Phys.Chem.B 1997; 101: 3460-3463.

[37]   Nagy J.B. Multinuclear NMR characterization of CTAB-hexanol-water sodium oleate –butanol –water and triton X 100 decanol water. Colloids and Surf.1989; 36:229-234.

[38]   Cason JP, Miller ME, Thompson JB, Roberts CB. Solvent effects on copper nanoparticle growth behavior in AOT reverse micelle system. J Phys Chem. B 2001; 105:2297-2302.

[39]   Pileni MP. Mesostructured fluids in oil rich Regions: Structural and templating approaches Langmuir 2001; 17: 7476.-7480.

[40]   Petit C, Wang Z L, Pileni M P. Seven-nanometer hexagonal close packed cobalt nanocrystals for high-temperature magnetic applications through a novel annealing process. J phys Chem B 2005; 109: 15309-15316.

[41]   Zhang DE, Ni XM, Zheng HG, Li Y, Zhang XI,Yang ZP. Synthesis of needle like nickel particles in water in oil microemulsions. Mater. Lett.2005; 59:2011-2014.

[42]   Khiew PS, Radiman S, Huang NM, Ahmad Md Soot, Nadarajah K.. Preparation and characterization of ZnS nanoparticles synthesized from chitosan laurate micellar solution. Mater Lett 2005; 59:989-993.

[43]   Petit C, Pileni MP Synthesis of cadmium sulfide in situ in reverse micelles and in hydrocarbon gels. J Phys Chem 1988; 92:2282-2285.

[44]   Estoe J, Cox AR. Formation of PbS nannocrystals using reverse micelles of lead and sodium AOT. Colloids Surf A Physicochem. Eng. Asp.1995; 101:63-76.

[45]   Ingerd D, Pileni MP. Limitations in producing nanocrystals using reverse micelles as nonreactors Adv. Funct Mater.2001; 11:136-139.

[46]   Arriagada FJ, Osseo-Assare K. Synthesis of nanosized silica in AOT microemulsions. J Colloid Interface Sci 1995;170: 8-17.

[7]   Lopez-Perez JA, Lopez –Quintela MA, Mira J, Rivas J, Charles SW , Advances in the Preparation of Magnetic Nanoparticles by the Microemulsion Method. J Phys. Chem. B 1997; 101, 8045-8047.

[48]   Nanni A, Dei L. Ca $(OH)_2$ nanoparticles from W/O microemulsions Langmuir 2003;19: 933-938.

[49]   Hota G, Idage S B, Khilar K C. Characterization of CdS-Ag_2S nanoparticles using XPS technique Colloids Surf A: Physiological and Engineering Aspects 2007; 293: 5-12.

[50]   Hota G Studies on the preparation of core-shell /composite nanoparticles using W/O microemulsions. Ph.D Thesis Indian Institute of Technology Bombay: 2004.

[51]   Kumar AR, Hota G, Mehra A, Khilar K C, Modeling of formation of nanoparticles by mixing two reactive microemulsions..AIChE J 2004; 50: 1556-1567.

[52]   Bandopadhyaya R, Kumar R, Gandhi K S Simulation of precipitation reactions in reverse micelles. Langmuir 2000; 16: 7139-7149.

[53]   Ethayaraja M, Dutta K, Muthukumaran D, Bandopadhyay R. Nanoparticle formation in water in oil microemulsions: Experiments, Mechanism and Monte Carlo simulation. Langmuir 2007; 128: 17102-17113.

[54]   Shukla D, Mehra A Modeling the formation of Shell in Core-Shell nanocrystals in reverse micellar system.Langmuir2006; 22: 9500-9506.

[55]   Viswanath B, Tikku S, Khilar K C. Modeling Core-shell nanoparticle formation using three reactive microemulsions. Colloids and Surfaces A 2007; 298: 149-157.

[56]   Sarkar D, Gupta P, Gautam A, Khilar KC Reuse of Surfactant/ Oil phase in Nanoparticle synthesis using W/O microemulsions.AIChE J 2008; 54 : 582-587.

[57]   Murphy CJ, Jana NRControlling the Aspect Ratio of Inorganic Nanorods and Nanowires. Adv. Mater.2002; 14: 80-82.

[58]   Kwan S, Kim F, Akana J, Yang P Synthesis and assembly of BaWO₄ nanorods. Chem. Commun. 2001; 447–448.

[59]   Yu, Chang S S, Lee C L, Wang CRC .Gold Nanorods: Electrochemical Synthesis and Optical Properties.J. Phys. Chem. B 1997: 101 (34): 6661-6664.

[60]   Huang L, Wang H, Wang Z, Mitra A, Bozhilov K N., Yan Y. Nanowire Arrays Electrodeposited from Liquid Crystalline Phases Adv. Mater.2002 ; 14: 61-64.

[61]   Esumi K, Matsuhisa K, Torigoe K, Preparation of Rodlike Gold Particles by UV Irradiation Using Cationic Micelles as a Template.Langmuir 1995; 11 : 3285-3287.

[62]   Kameo A., Suzuki A, Torigoe K, Esumi K Fiber-like Gold Particles Prepared in Cationic Micelles by UV Irradiation: Effect of Alkyl Chain Length of Cationic Surfactant on Particle Size. Journal of Colloid and Interface Science 2001; 241, 289–292.

[63]   Li M, Schnablegger H, Mann S Coupled synthesis and self-assembly of nanoparticles to give structures with controlled organization. Nature 1999; 402: 393-395.

[64]   Li Y, Li X, ,Deng Z X, Zhou B, Fan S, Wang J, Sun X From Surfactant-Inorganic mesostructures to tungsten nanowires. Angew. Chem. Int. Ed. 2002; 41:333-335.

CHAPTER 4

SYNTHESIS OF METAL NANOSTRUCTURES BY PHOTOREDUCTION

Bo Hu, Hong-Yan Shi and Shu-Hong Yu[*]

Division of Nanomaterials & Chemistry, Hefei National Laboratory for Physical Sciences at Microscale, the School of Chemistry & Materials, University of Science and Technology of China, P. R. China.

Address correspondence to: Shu-Hong Yu, Division of Nanomaterials & Chemistry, Hefei National Laboratory for Physical Sciences at Microscale, the School of Chemistry & Materials, University of Science and Technology of China, P. R. China; Tel: +86 551 3603040; E-mail: shyu@ustc.edu.cn

Abstract: Metal nanostructures with the size range of 1-100 nm, featuring unique physical and chemical properties that arise from their quantum size effects and high surface areas, have been the focus of recent scientific research. Among a variety of synthetic methods, the photoreduction method represents a promising strategy for controlled synthesis of metal nanostructures with different sizes, shapes and composition. In this review, the latest development on synthesis of metal nanostructures by UV radiation, γ-ray radiation, and laser radiation methods will be overviewed with specific examples to illustrate how to generate metal nanostructures with unusual structural specialty and complexity. The perspectives on combination of this method with solution processing and interfacial reactions are given.

Key words: Photoreduction, stabilization, polymers, UV-reduction, irradiation, nanoparticles.

1. INTRODUCTION

Recently, metal nanoparticles have attracted intensive scientific interest because of their unusual properties compared to the bulk metal (Ajayan Nature 1993) [1] (Ahmadi Science 1996) [2] (Andres Science 1996) [3] (Antonietti ACIE 1997) [4] (Antonietti ACIE 1997) [5]. Especially, nanoparticles of noble metals have been receiving much intensive research due to their potential applications in microelectronics (Antonietti ACIE 1997) [4] (Antonietti ACIE 1997) [5], optical (Guo AM 2008) [6] (Nehl JMC 2008) [7] (Tao Small 2008) [8], electronic (Teng JPCC 2008) [9] (Nishida Langmuir 2008) [10] (Didiot NN 2007) [11], and catalytic properties (Andres Science 1996) [3] (Moffitt CM 1995) [12] (Moffitt CM 1995) [13]. Current work of the synthesis of metal nanostructures mainly focused on several promising attempts: controlling over the shapes and sizes of noble metal nanoparticles (Ahmadi Science 1996) [2] (Antonietti ACIE 1997) [4] (Antonietti ACIE 1997) [5], producing one dimensional (1D) metal nanostructures(Foss JPC 1992) [14], synthesizing metal nanostructures-polymer nanocomposites (Nakao JCIS 1995) [15] (Zhu CC 1997) [16] (Yin CC 1998) [17], exploring the immobilization of metal nanoparticles on supports (Wang JSCC 1989) [18] (Ohtaki Macromolecules 1991) [19] (Toshima JMC 1994) [20] (Chen CC 1998) [21] (Chen AM 1998) [22], fabricating self-organized nanostructures from metal nanoparticles by polymeric and colloidochemical processes (Sarathy CC 1997) [23] (Selvan AM 1998) [24] (Selvan AM 1998) [25], and controlling self-assembly of metal nanostructures for the creation of structured nanocomposite materials (Shenhar AM 2005) [26] (Tang AM 2005) [27] (Yang JMC 2008) [28].

Some methods have been used in the recent years for the preparation of metal nanostructures. These methods generally involve the reduction of the relevant metal salts in the presence of a suitable surfactant, which is useful in the control of the growth of the metal particles. The photoreduction method, as a novel and facile technique, has demonstrated that it is possible to rationally design one-dimensional, two-dimensional, and even more complex superstructures of metal nanomaterials (Yu JNN 2004) [29] (Pillalamarri CM 2005) [30] (Amendola JPCB 2006) [31] (Zhang AFM 2007) [32]. This method can utilize hard or soft templates for the synthesis of unusual nanostructures (Esumi Langmuir 1995) [33] (Chang Langmuir 1999) [34]. Until now, this method has become more and more mature, and the synthesis scope have successfully extended to the monodisperse colloidal particles, shaped nanostructures, metal-inorganic nanocomposites, metal-polymer nanocomposites, and metal superstructures (Wang JSCC 1989) [18] (Toshima JMC 1994) [20] (Sarathy CC 1997) [23] (Yu JNN 2004) [29] (Curtis ACIE 1988) [35].

The photoreduction method can effectively control the size and shape of produced metal nanostructures. For example, UV photoreduction method has synthesized thin platelet-like Au nanocrystals with triangular or trancated hexagonal shape, the size and homogeneity of which also crucially depends on the type of the protective polymer (Mayer CPS 1998) [36]. The morphology of colloidal Au has been studied before (Milligan JACS 1964) [37], and shape-controlled synthesis of Au particles forming platelike trigons has been explained in terms of the Kossel-Stranski theory of face-selective growth of crystals. Shape-controlled synthesis of polymer protected Cu particles exhibiting platelike hexagonal morphology has also been reported (Curtis ACIE 1988) [35].

Several kinds of metal nanoparticles-polymer nanocomposites were reported. Noble metal solid sols in nonconducting system such as poly(methyl methacrylate) were prepared by polymerization of solutions of starting noble metal compounds in methyl methacrylate monomer and post-heating of the resulting solid solutions (Nakao JCIS 1995) [15]. γ-radiation method was applied for synthesis of polyacrylamide-silver (Zhu CC 1997) [16] and poly(butyl acrylate-*co*-styrene)-silver composites (Yin CC 1998) [17].

Immobilization of metal nanoparticles on supports seems to be a condition for these materials to retain high activity on recycling. Hirai and co-workers reported the immobilization of ultrafine rhodium particles on a polyacrylamide gel by forming an amide bond between the primary amino group of the support and the methylacrylate residue in the protective polymer (Ohtaki Macromolecules 1991) [19]. Liu *et al.* investigated the capture of colloidal metal particles on the surface of functionalized silica by ligand coordination (Wang JSCC 1989) [18]. Akashi and co-workers reported the synthesis of Pt nanpoparticles on polystyrene micorspheres, including the reduction of H_2PtCl_6 by aqueous ethanol in the presence of polystyrene microspheres with surface-grafted poly (*N*-isopropylacrylamide) (Chen CC 1998) [21]. Later, they further developed the in-situ synthesis of Ag nanparticles on poly (*N*-isopropylacrylamide)-coated polystyrene microspheres using 2,2'-azobisisobutyronitrile as initiator of polymerization (Chen AM 1998) [22].

The photoreduction method has successfully fabricated the metal superstructures. For instance, Rao *et al.* (Sarathy CC 1997) [23] (Sarathy JPCB 1997) [38] reported a novel method of preparing thiol-derivatised nanoparticles of Au, Pd, and Ag forming superstructures. Thio-derivatised nanoparticles of Au, Pd, and Ag forming superstructures are recently prepared by the acid-facilitated transfer of well characterized particles in hydrosol to a toluene layer containing the thio. Recently, Selvan *et al.* (Selvan AM 1998) [24] (Selvan, Spatz, Klok and Moller 1998) [25] reported a method for producing gold-polypyrrole composites by using a micellar solution of polystyrene-block-poly(2-vinylpyridine) (PS-P2VP) treated with tetrachloroauric acid and the polymerization of pyrrole (PY). The gold-PY core-shell particles in PS-P2VP (Selvan AM 1998) [25] and

a novel elegant dentritic supramolecular nanostructure are observed (Selvan AM 1998) [24].

The photoreduction method for preparation of metal nanostructures was established by Henglein *et al.* (Mosseri JPC 1989) [39] and Fendler *et al.* (Kurihara JACS 1983) [40]. Compared with the chemical reduction of metal ions, this photoreduction route has important advantages: it is reproducible and can be applied at ambient temperature, no disturbing chemical impurities are introduced, and the reduction is initiated homogeneously without local concentration gradients when reactants are mixed. In the development process of photoreduction method, three fundamental methods, which are UV radiation, γ-Ray radiation, and laser radiation methods, have emerged and attracted great research attention. Although three fundamental photoreduction methods have different light source and reaction mechanism, they have been widely employed for producing colloidal metal nanoparticles, shaped metal nanostructures, organic-metal nanocomposites, and metal superstructures.

In this chapter, a general overview on the recent development in the area of the photoreduction method to the synthesis of metal nanostructures with structural complexity and specialty will be given. We will carefully carry out our discussion of three fundamental photoreduction methods, especially focusing on the reaction mechanism, current situation and some elegant examples. Finally, we will present the summary and outlook of the photoreduction method to the synthesis of metal nanostructures, and conceive the challenges that must be confronted before their applications.

2. SYNTHESIS OF METAL NANOSTRUC-TURES BY PHOTOREDUCTION METHOD

The photoreduction method is devoted to metal nanostructures synthesis in solution by the radiolytic reduction of ionic precursors. The specificity of the radiation-induced reduction process is to generate radiolytic radicals of strongly reducing potential, which is more negative than that of any ion. The free ions are generally reduced at each encounter, so that the individually formed atoms coalesce and yield monodisperse clusters. While the excess metal ions adsorb at the surface of these clusters, they can also be reduced by radiolytic radicals. The radiolytic radicals can greatly influence the nucleation and growth process of metal nanostructures. The radiation can homogenously penetrate into pores or the surface of complex supports, forming the clusters in situ. The reduction process is performed at room temperature, so that supports fragile to heat may be used in the radiation process. Further, the reduction rate can be controlled by the selected dose rate, showing a wide range of conditions from quasi-instantaneous to slow atom production.

In the radiation process, while energy deposit throughout the solution, the radiolytic radicals are formed with the initial homogeneous distribution by ionization and excitation of the solvent. An elegant example is the aqueous solution: (Baxendale NATO ASI 1982) [41].

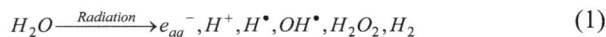

$$H_2O \xrightarrow{Radiation} e_{aq}^{-}, H^{+}, H^{\bullet}, OH^{\bullet}, H_2O_2, H_2 \qquad (1)$$

The solvated electrons e_{aq}^{-}, and $H^{\bullet}$ atoms are strong reducing agents $\left[E^{0}\left(H_2O / e_{aq}^{-} \right) = -2.87 \ V_{NHE} \right]$ and $\left[E^{0}\left(H^{+} / H \right) = -2.3 \ V_{NHE} \right]$, (Belloni NJC 1998) [42] so that they can easily reduce metal ions to the metal atom:

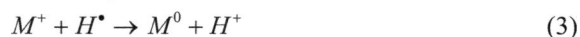

$$M^{+} + e_{aq}^{-} \rightarrow M^{0} \qquad (2)$$

$$M^{+} + H^{\bullet} \rightarrow M^{0} + H^{+} \qquad (3)$$

Where M^{+} are the monovalent metal ions. Further, multivalent ions are reduced by multistep reactions, including disproportionation of intermediate valencies. The reduction reactions are often diffusion-controlled. Moreover, the solution is usually added with a $OH^{\bullet}$ radical scavenger, because $OH^{\bullet}$ radicals can oxidize the ions or the atoms into a higher oxidation state and thus to counterbalance the above reduction reactions. The preferred scavenger is the solute molecules whose oxidation by $OH^{\bullet}$ produce radicals that not only are unable to oxidize the metal ions, but exhibit strong reducing ability, such as the radicals of secondary alcohols or of the formate anion:

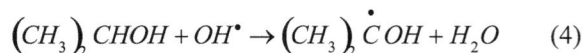

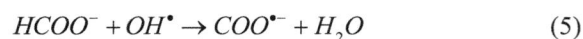

$$(CH_3)_2 CHOH + OH^{\bullet} \rightarrow (CH_3)_2 \overset{\bullet}{C} OH + H_2O \qquad (4)$$

$$HCOO^{-} + OH^{\bullet} \rightarrow COO^{\bullet-} + H_2O \qquad (5)$$

$H^{\bullet}$ radicals are scavenged by these molecules as well:

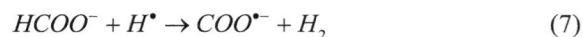

$$(CH_3)_2 CHOH + H^{\bullet} \rightarrow (CH_3)_2 \overset{\bullet}{C} OH + H_2 \qquad (6)$$

$$HCOO^{-} + H^{\bullet} \rightarrow COO^{\bullet-} + H_2 \qquad (7)$$

The radicals $(CH_3)_2 \overset{\bullet}{C} OH$ and $COO^{\bullet-}$ are as strong

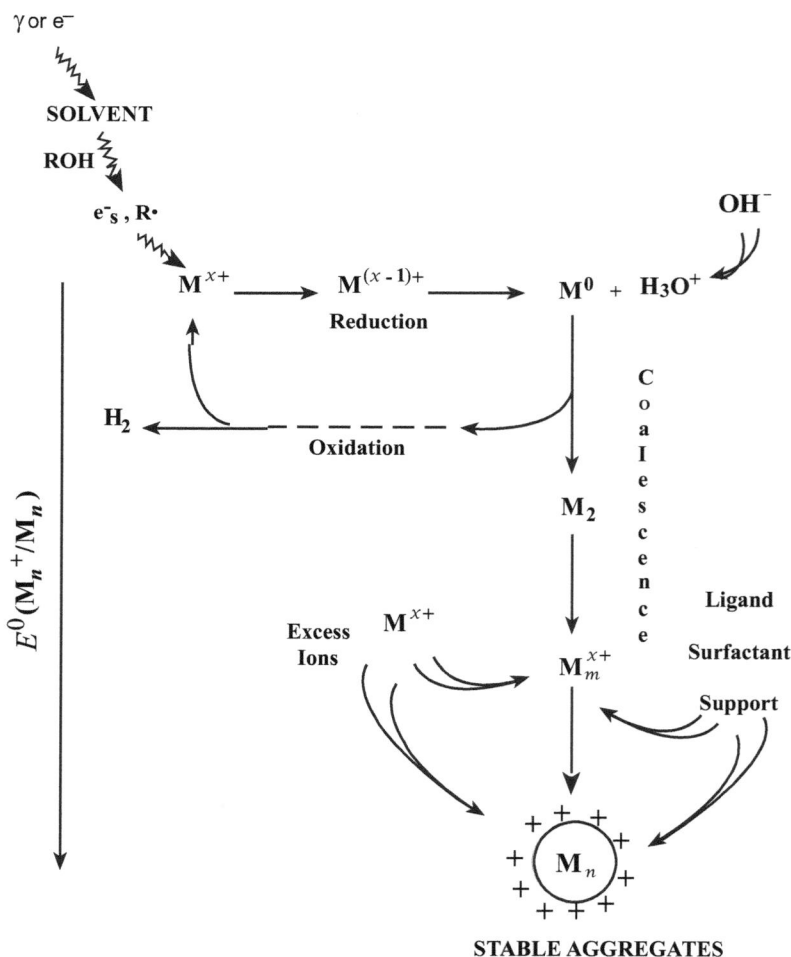

Fig. (1). Scheme of metal ion reduction in solution by ionizing radiation. The isolated atoms formed coalesce into clusters. They adsorb an excess of ions. They are stabilized by polymer, surfactant, small molecules, support, or some ligands. The redox potential increases with the nuclearity. Images reprinted with permission from Ref. (Belloni NJC 1998) [42]. Copyright © 1998 The Royal Society of Chemistry.

reducing agents as $H^\bullet$ atom:

$$E^0\left[\left(CH_3\right)_2 CO / \left(CH_3\right)_2 \overset{\bullet}{C} OH\right] = -1.8\ V_{NHE}\ \text{at pH=7}$$

and $E^0\left(CO_2 / COO^{\bullet-}\right) = -1.9\ V_{NHE}$, respectively. (Belloni, Mostafavi, Remita, Marignier and Delcourt 1998) [42] Importantly, based on these radicals, the reduction reactions are shown as follows:

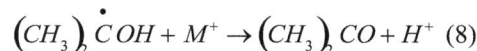

$$\left(CH_3\right)_2 \overset{\bullet}{C} OH + M^+ \rightarrow \left(CH_3\right)_2 CO + H^+ \quad (8)$$

$$COO^{\bullet-} + M^+ \rightarrow CO_2 + M^0 \quad (9)$$

The metal atoms are formed in a homogenous way. Because the binding energy between two metal atoms is stronger than the atom-solvent or atom-ligand bond energy, these metal atoms will coalesce when they encounter or associated with an excess of ions: (Belloni NJC 1998) [42].

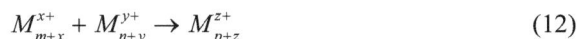

$$M^0 + M^0 \rightarrow M_2 \quad (10)$$

$$M^0 + M^+ \rightarrow M_2^+ \quad (11)$$

$$M_{m+x}^{x+} + M_{n+y}^{y+} \rightarrow M_{p+z}^{z+} \quad (12)$$

where m, n and p represent the nuclearities (i.e. the number of reduced atoms they contain) and x, y and z are the number of associated ions. If polymer, surfactant, small molecules, support, or some ligands are employed for stabilizing the radiation-induced small clusters and nuclear, these clusters and nuclear can be survived in the solution. (Belloni NJC 1998) [42] The metal ions reduction process is simply summarized in Fig. (**1**).

Further, radicals generated in a solution which contains nanoparticles can diffuse to the surface of particles and inject electrons or positive holes into them, changing the surface charge distribution. These processes often proceed at a diffusion controlled rate. They usually result in the changes of the electrical conductivity of the solution as H^+ ions are formed or consumed. For example, the 1-hydroxymethylethyl radicals react with large silver particles and reactions are shown as follows: (Henglein JPC 1993) [43].

$$x\left(CH_3\right)_2 COH + Ag_n \rightarrow x\left(CH_3\right)_2 CO + Ag_n^{x-} + xH^+ \quad (13)$$

The nanoparticles may obtain electrons from many radicals. By measuring the increase in conductivity, the number of stored electrons can be determined. Similarly, the hydroxyl radicals that are generated by irradiation can inject positive holes to the surface of particles, forming Ag_n^{x+} particles. Therefore, it is interesting to investigate the chemistry and the optical properties of the Ag_n particles carrying with excess Ag_n^{x+} and Ag_n^{x-} particles.

3. ULTRAVIOLET RADIATION METHOD

UV photoreduction method is the use of Ultraviolet light as light sources, in which Ultraviolet light is electromagnetic radiation with a wavelength range from 400 nm to 10 nm, and energies from 3 eV to 124 eV. Until now, UV photoreduction method has become a mature approach for the synthesis of various metal nanomaterials, especially including monodisperse clusters and nanoparticles (Yu JNN 2004) [29] (Itakura Langmuir 1995) [44] (Yonezawa JCSFT 1991) [45] shaped nanostructures, (Zhou AM 1999) [46] (Zhou CM 1999) [47], organic-metal nanocomposites (Harada JCIS 2008) [48] (Tan Langmuir 2007) [49] (Tamai Langmuir 2008) [50], and thin films (Wang Langmuir 2002) [51] (Joly Langmuir 2000) [52]. More and more metal nanomaterials, synthesized from UV photoreduction method, have been researched for the organization of complex structures and the fabrication of nanodevices.

UV photoreduction method has been demonstrated that it is an important approach for the synthesis monodisperse metal nanoparticles. For example, Toshima *et al.* (Toshima BCSJ 1992) [53] prepared colloidal Pt and Pd nanoclusters by UV photoreduction in the presence of surfactants. Then, this method is attractive in producing other monodisperse metal and bimetal nanoparticles (Yonezawa, Sato, Kuroda JCSFT 1991) [45] (Sato AOC 1991) [54]. This technique was developed further for producing metal nanoparticles in the polymer gel (Torigoe Langmuir 1993) [55] (Torigoe Langmuir 1992) [56].

A typical example is the synthesis of gold nanoparticles (AuNPs) with different sizes by UV irradiation-enhanced reduction in the presence of double hydrophilic block copolymers (DHBCs) with different functional patterns (Yu JNN 2004) [29]. The particle size could be controlled by variation of the functional group patterns of the block copolymers or UV irradiation power. While phosphonated poly(ethylene glycol)-block-poly(methacrylic acid)-PO$_3$H$_2$ (PEG-*b*-PMAA–PO$_3$H$_2$), poly(ethylene glycol)-block-poly[(*N*-carboxymethyl)ethyleneimine] (PEG-*b*-PEIPA), and poly(ethylene glycol)-block-poly(hydroxyl ethylene)-PO$_4$H$_2$ (PEG-*b*-PHEE–PO$_4$H$_2$) (30%) were separately added into the reaction system, the sizes of AuNPs were 4 nm, 28-35 nm, and 6 nm, respectively Fig. (**2a-c**). Using a carboxylated poly(ethylene glycol)-block-poly(methacrylic acid) (PEG-b-PMAA) polymer leaded to the formation of nonspherical nanoparticles with relatively broad size distribution Fig. (**2d**).

UV radiation technique has also demonstrated that it is possible to produce shaped metal nanomaterials. Recently, Esumi and his co-workers (Esumi Langmuir 1995) [33] successfully obtained rodlike Au nanoparticles by UV irradiation using rodlike

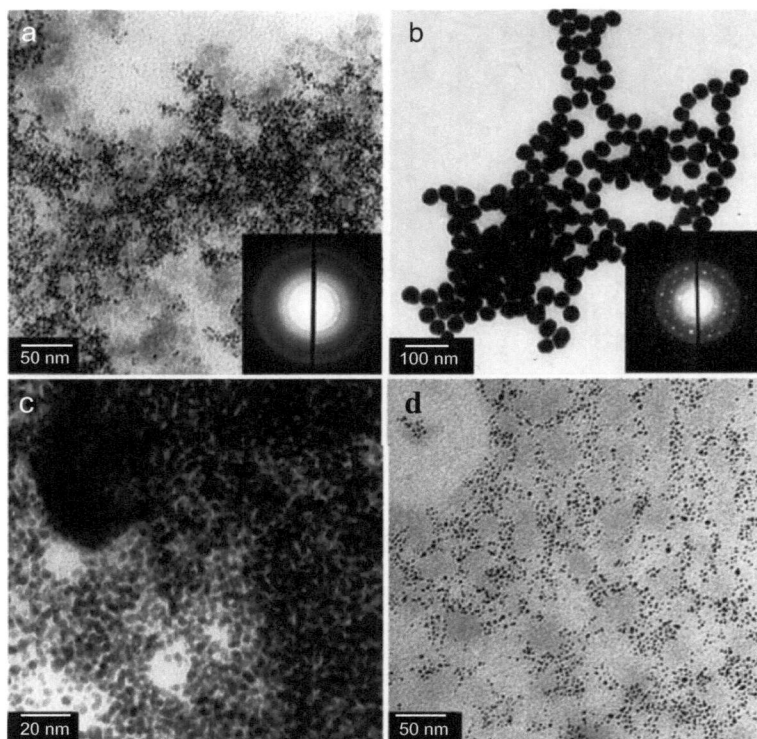

Fig. (2). TEM images of Au nanoparticles synthesized by UV irradiationenhanced reduction of 10^{-4} M HAuCl$_4$ solution at 200 W, [polymer] = 1 g/L. (**a**) PEG-*b*-PMAA–PO$_3$H$_2$; (**b**) PEG-*b*-PEIPA; (**c**) PEG-*b*-PHEE–PO$_4$H$_2$ (30%); (**d**) PEG-*b*-PMAA. Images reprinted with permission from Ref. (Yu JNN 2004) [29] Copyright © 2004 American Scientific Publishers.

hexadecyltrimethylammonium chloride (HTAC) micelles as soft template. However, the rodlike Au nanoparticles are not so uniform. Nonspherical Pd nanocrystals in SDS/poly (acrylamide) gel were also obtained by UV irradiation (Torigoe Langmuir 1995) [57]. Another interesting class of colloids generated by UV-reduction are large, but thin platelet-like Au nanocrystals with triangular or trancated hexagonal shape, the size and homogeneity of which also crucially depends on the type of the protective polymer (Mayer CPS 1998) [36] (Mayer APACS 1998) [58].

Zhou *et al.* (Zhou AM 1999) [46] have reported the shape-controlled synthesis of single-crystal silver nanostructures via the UV irradiation photoreduction technique at room temperature using polyvinyl alcohol (PVA) as the protecting agents. Ag nanorods about 15-20 nm in diameter and up to 350 nm in length were synthesized Fig. (**3a**). Increasing the concentration of AgNO$_3$ in the solution, the Ag nanorods grew thicker and longer Fig. (**3b**). With the further increasing the concentration of AgNO$_3$ in the system, elegant Ag dendrites were formed Fig. (**3c**). Therefore, the concentration of AgNO$_3$ played a significant role in the formation and growth of Ag nanostructures.

Moreover, zhou *et al.* (Zhou CM 1999) [47] have reported the shape-controlled synthesis of AuNPs using the UV irradiation technique at room temperature. The slow reduction rate of the UV irradiation process may favor the nucleation and

growth of shaped gold nanostructures. In a typical preparation, platelike gold triangles with the 15 nm size were synthesized Fig. (**3d**). Some experimental parameters played important roles in the morphology control of AuNPs: increasing the concentration of gold cations, these platelike nanostructures grew larger and to develop into hexagonal shapes Fig. (**3e**); prolonging the irradiation time, AuNPs with more regular shape were formed; increasing the concentration of polyvinyl alcohol (PVA), AuNPs with polyhedral shapes were produced; displacing PVA with poly(ethylene glycol) (PEG), AuNPs with quasiellipsoidal morphology were synthesized Fig. (**3f**).

The combination of the UV photoreduction methods with mesoporous hard templates such as silica (Sakamoto JPCB 2004) [59] (Fukuoka MMM 2001) [60], alumina (Yu CC 2000) [61] (Foss JPC 1992) [62], and carbon nanotube (Grobert CC 2001) [63] (Govindaraj CM 2000) [64] (Sloan CC 1998) [65], allows it possible to synthesize metal nanowires. Especially mesoporous silica such as FSM-16 (Yanagisawa BCSJ 1990) [66] (Inagaki JCSCC 1993) [67], MCM-41 (Kresge Nature 1992) [68], and SAB-15 (Zhao Science 1998) [69] are very suitable for the synthesis of metal nanostructures, due to large pores (2-50 nm) with the narrow distribution and high surface area (up to 1000 m^2 g^{-1}). These mesoporous silica have been widely used as hard templates to synthesize metal nanowires in the one-dimensional channel structures. Fukuoka *et al.* (Fukuoka ICA 2003) [70] (Fukuoka JACS 2001) [71] have successfully

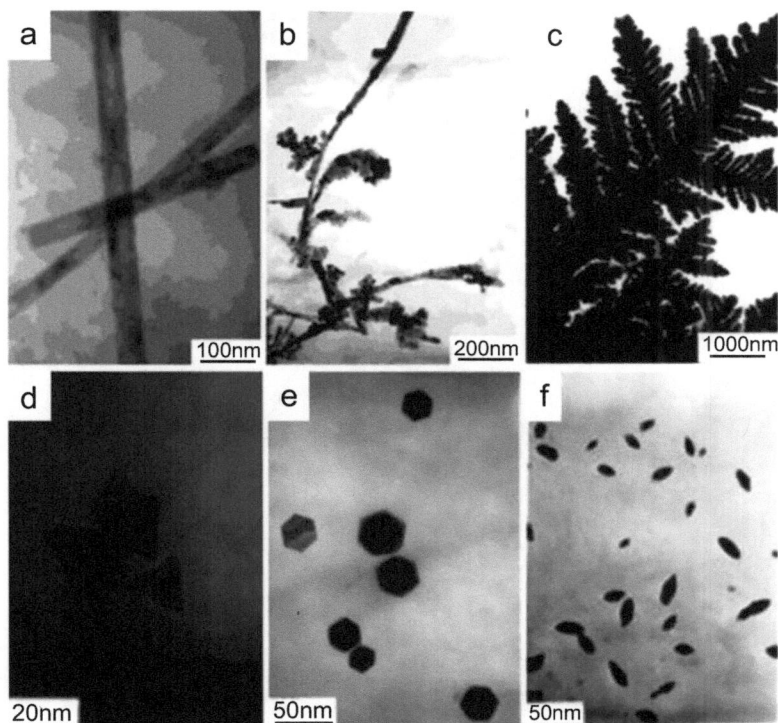

Fig. (3). (a) TEM image of the product by irradiating the solution containing 10^{-4} M AgNO$_3$. **(b)** TEM image of the product obtained by irradiating the solution containing 10^{-3} M AgNO$_3$. **(c)** TEM image of the product obtained by irradiating the solution containing 10^{-2} M AgNO$_3$. **(d)** TEM image of the platelike triangular gold nanoparticles, **(e)** the hexagonal-shaped gold nanoparticles, **(f)** the quasi-ellipsoidal shaped gold nanoparticles. Images **(a)**-**(c)** reprinted with permission from Ref. (Zhou AM 1999) [46] Copyright © 1999 Wiley-VCH. Images d)-f) reprinted with permission from Ref. (Zhou CM 1999) [47] Copyright © 1999 The American Chemical Society.

reported the synthesis of some mono- and bimetallic nanowires of Pt, Rh, Au, Pd, Pt-Rh, and Pt-Pd in HMM-1 containing ethane fragments via the photoreduction method.

Ichikawa *et al.* have demonstrated that Pt nanowires can be synthesized in HMM-1 templates with high yields by the UV photoreduction methods (Sakamoto JPCB 2004) [59]. UV radiation of the sample for 48 h led to the formation of Pt nanowires with the diameter of 3 nm and the length ranges from 10 to several hundred nanometers in HMM-1 templates Fig. (**4a**) and (**b**). On the other hand, H$_2$-reduction of the sample resulted in the formation of Pt nanoparticles (diameter 3 nm) and short nanowires in the channels Fig. (**4c**). These differences were influenced by the rate of reduction and migration of Pt ions. If the migration rate of Pt ions in the water/methanol phase was slower than the reduction rate, they could be catalyticlly reduced on Pt particles to grow to wires. On the contrary, in the H$_2$-reduction, they resulted in the formation of nanoparticles. The isolation of Pt nanowires was possible by removing HMM-1 with HF, and these nanowires could be stabilized by modification with ligands, such as [N(C$_{18}$H$_{37}$)(CH$_3$)$_3$]Cl and P(C$_6$H$_5$)$_3$ Fig. (**4d-f**). The ligand-modified nanowires were well dispersed compared to the surfactant-free nanowires.

The mechanism for the formation of the Pt nanowires via the UV photoreduction method was very interesting. First, Pt ions could be reduced, forming tiny Pt nanoparticles in the mesoporous channels. Because organic radicals, such as $^{\bullet}CH_2OH$ were produced from adsorbed methanol by UV irradiation, had strong reduction ability that could easily reduce Pt salts. Then, the soluble Pt ions in the water/methanol phase could migrate in the channels to the surface of Pt nanoparticles. These Pt ions were catalytically reduced on the surface, and the gradual growth of Pt nanoparticles resulted in nanowires. Therefore, the promotion of the migration of Pt ions and small Pt species in the one-dimensional channel was the key to synthesize the Pt nanowires.

A variety of soft templates have been employed in the preparation of metal nanostructures with controlled size and spatial distribution. Some soft materials such as peptides (Banerjee PNAS 2003) [72], organic polymers, biopolymers (Sun ASS 2006) [73] (Aldaye Science 2008) [74], and biological systems (Flynn AM 2003) [75] can be used as a template on which nanostructures can be synthesized and assembled. Further, a variety of organic matrix made from gels (Antonietti CEJ 2004) [76] and polymers (Zhong Langmuir 2008) [77] can be employed for the stabilization and formation of the metal nanostructures.

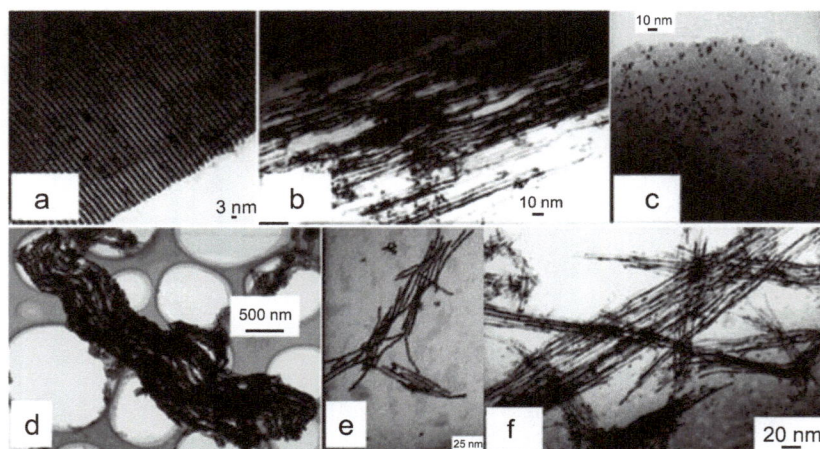

Fig. (4). TEM images of (**a**) HMM-1, (**b**) Pt wire/HMM-1 synthesized by photoreduction, and (**c**) Pt particle/HMM-1 synthesized by H$_2$-reduction. TEM images of unsupported Pt wires: (**d**) ligand-free Pt wires; (**e**) [N(C$_{18}$H$_{37}$)-(CH$_3$)$_3$]Cl-modified Pt wires; (**f**) P(C$_6$H$_5$)$_3$-modified Pt wires. Images reprinted with permission from Ref. (Sakamoto JPCB 2004) [59]. Copyright © 2004 The American Chemical Society.

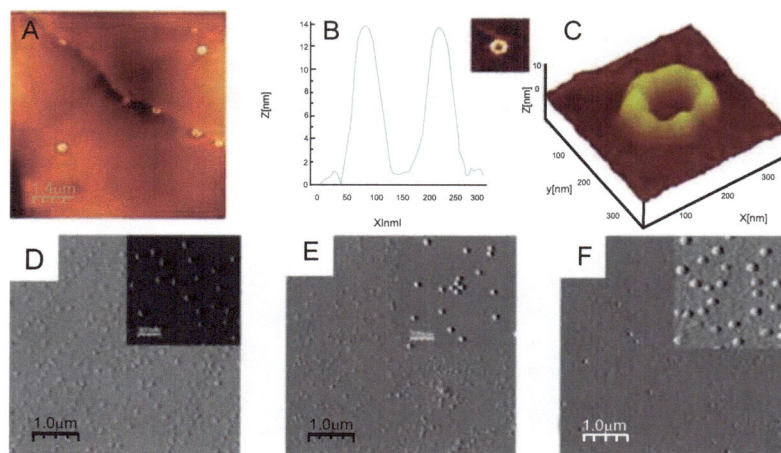

Fig. (5). (**A**) AFM height image of the toroidal topology of plasmid pcDNA 3.1(+) on a HOPG substrate. (**B**) Height analysis of inset (top right corner). (**C**) 3-Dimensional AFM image of the plasmid shown in B. AFM amplitude images of metal nanoparticles obtained from photo-initiated reduction of metal ions bound to plasmid pcDNA 3.1(+) sacrificial mold: (**D**) gold; (**E**) nickel; (F) cobalt particles. Images reprinted with permission from Ref. (Samson ACSN 2009) [78] Copyright © 2009 The American Chemical Society.

The plasmid is the circular extra-chromosomal DNA molecule, which is capable of replicating autonomously in bacterial hosts and can be produced in large quantities. Drain *et al.* (Samson ACSN 2009) [78] have summarized several advantages of plasmid DNA as the soft templates: (1) DNA is well-known to bind cations at the negatively charged phosphate backbone with various affinities, (2) metal ion binding favors formation of toroidal plasmid DNA structures, (3) the size of plasmid, therefore the size of the toroid, is easy to vary, (4) DNA has a well-established UV light initiated oxidation chemistry.

Moreover, using the templates of the biological sacrificial toroidal plasmid DNA, Drain *et al.* (Samson ACSN 2009) [78] had successfully synthesized narrowly dispersed gold, nickel, and cobalt metal nanodiscs at room temperature with UV irradiation Fig.

(**5**). The toroidal DNA with different numbers of base pairs and sequences could control the formation process and size of the nanostructures. UV irradiation could initiate the oxidation of the DNA molecules and accompany the reduction of the DNA bound metal ions. This method is general and may be applicable to DNA/RNA structures with complex shape for the formation of novel metal nanostructures.

Bimetal nanostructures are of interest from both technological and scientific points of view for improving catalytic activity and optical properties (Itakura Langmuir 1995) [44] (Link JPCB 1999) [79] (Treguer JPCB 1998) [80] (deCointet JPCB 1997) [81] (Schmid ACIE 1991) [82]. There are generally two ways for the preparation of bimetallic nanoparticles from metal salts. One is the coreduction method that is used in the preparation of monometallic nanoparticles

(Itakura Langmuir 1995) [44] (Link JPCB 1999) [79]. The other is the successive reduction method that is usually carried out to prepare core-shell nanostructures (Treguer JPCB 1998) [80] (deCointet JPCB 1997) [81] (Schmid ACIE 1991) [82]. Since Pal *et al.* (Mallik NL 2001) [83] reported for the first time a photochemical method for the preparation of core-shell bimetallic nanoparticles in 2001, the UV irradiation method has developed and played an important role in the design and synthesis of them. In this work, a novel and reproducible photochemical technique for the preparation and growth of gold-silver bimetallic core-shell nanoparticles has been reported (Mallik NL 2001) [83]. The formation mechanism included a two-step process. In the first step, gold ions were reduced in solution, and the produced atoms agglomerated to form small clusters, then the formed nanoclusters acted as nucleation centers and catalyzed the reduction process of the remaining gold ions present in the bulk [83]. In the second step, gold nanoparticles served as templates, and silver ions (M^{*+}) adsorbed on the surface of the gold (M) nanoparticles and reduced by the suitable conditions, forming monolayer of silver similar to $M_{core} - M^{*}_{shell}$ structures. A series of M^{*} layers formed on the surface of gold seeds, and the sizes of final bimetallic nanoparticles could be controlled by the ratio of seeds to the metal ions [83].

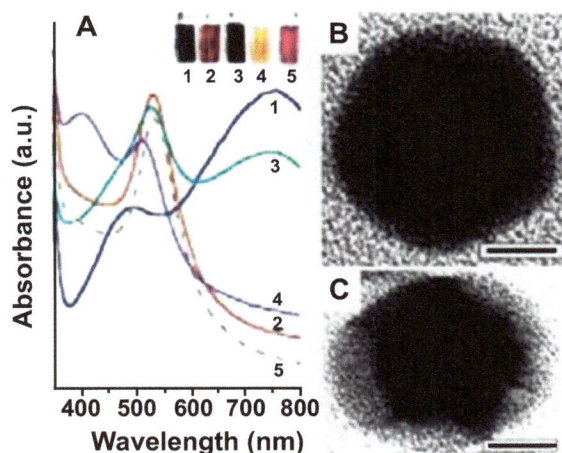

Fig. (6). (A) UV-vis spectra recorded from (1) 10^{-2} M aqueous solution of PTA after UV irradiation; (2) UV-irradiated PTA solution after addition of 10^{-3} M HAuCl$_4$; (3) solution 2 after further UV irradiation; (4) solution 3 after addition of 10^{-3} M Ag$_2$SO$_4$ solution; and (5) solution 2 after addition of 10^{-3} M Ag$_2$SO$_4$. Pictures of sample bottles containing solutions 1-5 are shown in the inset. **(B)** TEM image of one of the gold particles, and **(C)** the Au core-Ag shell particles. Scale bars in C and D correspond to 5 and 10 nm, respectively. Images reprinted with permission from Ref. (Mandal JACS 2003) [84] Copyright © 2003 The American Chemical Society.

Recently, some smart strategies have been designed for the improvement of the synthesis of core-shell bimetallic nanoparticles. These strategies include the immobilization of a reducing agent on the surface of the seed metal nanoparticles and reaction with the second metal ions, then these ions will be reduced on the surface, forming a thin metallic shell. Based on these strategies, Sastry *et al.* (Mandal JACS 2003) [84] have used Keggin ions as UV-switchable reducing agents in the synthesis of gold-silver bimetallic core-shell nanoparticles Fig. (**6**). TEM images had shown that the core-shell nanoparticles had a distinct variation in contrast between the dark gold core and the lighter silver shell Fig. (**6D**). In the synthesis process, photochemical charging of Keggin ions, such as phosphotungstic acid (PTA) molecules, bound to AuNPs was the crucial step. The variation of the reducing capability of the Keggin ions using UV irradiation was an additional feature that could enhance the versatility of this strategy. Therefore, it is great potential to develop this new strategy for realizing bimetallic core-shell structures using Keggin ions with potential application in nanomaterials synthesis.

Hybrid polymer-metal materials by immobilizing metal nanostructures on the surface or inner space of polymer materials have been studied intensively (Chen CM 1999) [85] (Dokoutchaev CM 1999) [86] (Kim JACS 2002) [87] (Korchev Langmuir 2006) [88] (Li Langmuir 2002) [89] (Lu ACIE 2006) [90] (Mayer JPCB 2000) [91] (Mayer AMC 1999) [92] (Mei CM 2007) [93] (Mei Langmuir 2005) [94] (Pathak CM 2000) [95] (Sakamoto CM 2008) [96] (Schuetz CM 2004) [97] (Shi Langmuir 2005) [98] (Suzuki Langmuir 2005) [99] (Wen CM 2008) [100]. There are generally two ways for the production of thESE hybrid materials. One is the direct adsorption of metal nanostructures on the surface of polymer materials that have functional groups such as thiol and pyridine groups (Dokoutchaev CM 1999) [86] (Pathak CM 2000) [95] (Shi Langmuir 2005) [98]. The other is the in situ reduction of metal ions forming metal nanostructures on the surface or inner space of polymer materials (Suzuki Langmuir 2005) [99] (Wen CM 2008) [100]. These hybrid materials have shown the collectively unusual properties of metal nanostructures and polymer supports, and great potentials in the catalytic, imaging, and sensing applications (Li, Boone and El-Sayed 2002) [89] (Shi Langmuir 2005) [98] (Suzuki Langmuir 2005) [99] (Wen CM 2008) [100].

An elegant example is the synthesis of polymer-metal (Au, Ag, and Pd) hybrid materials via the UV irradiation of the the polystyrene particles incorporating PMPS in the presence of metal salts Fig. (7) (Tamai Langmuir 2008) [50]. The reduction of metal ions was accompanied with the UV photoinduced oxidation of PMPS. Metal nanoparticles nucleated and grew on the polymer particle surface, forming hybrid materials. Some parameters affected the formation and growth of the metal nanoparticles. First, the charge of the polymer particles and the metal ions had the strong influence role. For example, the

UV irradiation of the cationic polymer particles with the cationic metal ion (AgNO$_3$ and PdCl$_2$) did not result in the formation of metal nanoparticles, in contrast to the case of anionic polymer particles. Further, the functional surface groups of polymer particles could affect the shape of the palladium nanoparticles. The functional groups on the surface of the anionic polymer particles, such as *N*-isopropylamide and acetoacetoxy groups, could facilitate the formation of the larger shapeless palladium nanoparticles.

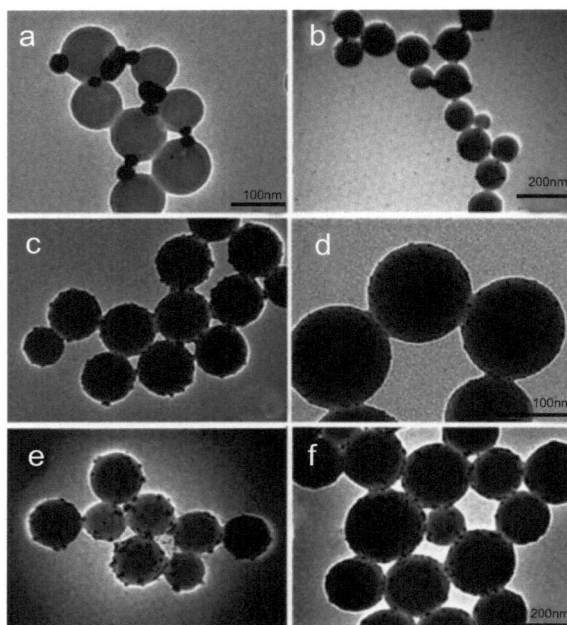

Fig. (7). TEM images of the polymer-metal nanoparticles hybrid particles: (**a**) 2a/HAuCl$_4$ ·4H$_2$O and (**b**) 1a/AgNO$_3$. TEM images of the polymer- palladium nanoparticle hybrid particles: (**c**) 2a/Na$_2$PdCl$_4$, (**d**) 2b/Na$_2$PdCl$_4$, (**e**) 3a/Na$_2$PdCl$_4$, and (f) 2a/PdCl$_2$. Images reprinted with permission from Ref. (Tamai Langmuir 2008) [50] Copyright © 2008 The American Chemical Society.

Recently, semiconductor-metal nanocomposites have been intensively investigated, which have some important applications such as biolabels, electroluminescent displays, memory devices, photochemical solar cells, and sensors (Bruchez Science 1998) [101] (Huynh Science 2002) [102]. A novel site-specific photodeposition method has been designed for the formation of semiconductor-metal nanocomposites (Pacholski ACIE 2004) [103] (Dukovic AM 2008) [104] (Cozzoli JACS 2004) [105] (Ohtani JPCB 1997) [106] (Ng AM 2007) [107]. The reaction mechanism includes that irradiation of the solution of semiconductor nanostructures containing metal ions leads to the formation of metal nanoparticels by the reduction role of the conduction band electrons, and concomitantly to an oxidation of the hole scavengers by valence band holes.

Weller *et al.* have successfully synthesized a nanoheterostructure consisted of a ZnO nanorod with

an attached spherical Ag particle at one end by the anisotropic photoreduction of silver ions at the ZnO (Pacholski ACIE 2004) [103]. UV irradiation of the solution of ZnO nanoparticles in ethanediol/water containing silver ions leads to silver nanoparticle formation by the reduction of the conduction band electrons, and concomitantly to an oxidation of the alcohol by valence band holes. The silver nanoparticles usually located at the end of the ZnO nanorods Fig. (**8**), because that the photoreduction preferentially occurs at one end of the ZnO nanorods, which favored the nucleation and growth of silver nanoparticles at this specific site. The further development of this site-specific photodeposition concept could lead to the custom-made construction of electronic and optical devices.

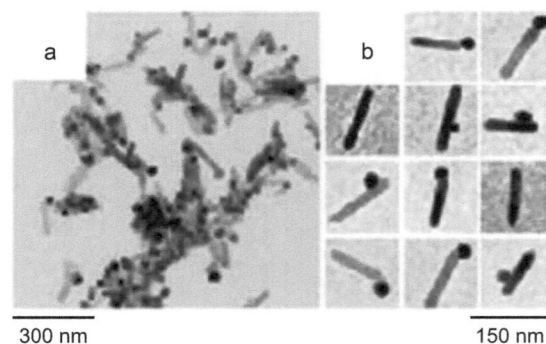

Fig. (8). TEM images of Ag$^+$-containing ZnO nanorod solutions after irradiation. (**a**) ZnO nanorods with deposited silver particles. (**b**) Irradiation experiment with poly(ethylenimine) as stabilizer. Images reprinted with permission from Ref. (Pacholski ACIE 2004) [103] Copyright © 2004 Wiley-VCH.

4. γ-RAY RADIATION METHOD

γ-ray radiation from ^{60}Co source is often used for the preparation of small colloid particles and the charging of particles by excess electrons or positive holes. In the colloid preparation, the hydrated electrons and reducing organic radicals that are generated in the radiolysis of aqueous solutions reduce dissolved metal ions, thus producing unusual valency states. The reaction in the solution can be expressed as following equations: (Marignier Nature 1985) [108].

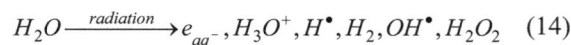

$$H_2O \xrightarrow{radiation} e_{aq^-}, H_3O^+, H^\bullet, H_2, OH^\bullet, H_2O_2 \quad (14)$$

$$e_{aq^-} + M^{m+} \rightarrow M^{(m-1)+} \quad (15)$$

$$e_{aq^-} + M^+ \rightarrow M^0 \quad (16)$$

$$nM^0 \rightarrow \cdots\cdots M_n \cdots\cdots \rightarrow M_{agg} \quad (17)$$

For example, silver ions are reduced to yield free atoms in solution which subsequently coalesce to form larger particles (Henglein JPC 1993) [43]:

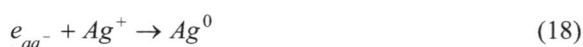

$$e_{aq^-} + Ag^+ \rightarrow Ag^0 \quad (18)$$

$$nAg^0 \rightarrow Ag_n \qquad (19)$$

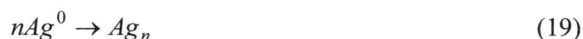

Polymers in small concentration, such as polyphosphate, polyacrylate, poly-(vinyl sulfate), poly-(vinyl alcohol), and poly(ethyleneimine), are often used to stabilizer the growing aggregates (Henglein JPC 1993) [43]. γ-radiation method has been widely applied to reduce many metal ions for producing metal nanoparticles, bimetallic particles and alloy (Marignier Nature 1985) [108] (Zhu ML 1993) [109] (Zhu JMC 1994) [110] (Zhu MSEBS 1994) [111]. Now, it has been extended to synthesize various materials such as metal-inorganic nanocomposites (Zhu ML 1996) [112] and metal-polymer nanocomposites (Zhu CC 1997) [16].

An elegant example is the radiolytic control of the size of colloidal gold nanoparticles Fig. (**9**) (Henglein Langmuir 1998) [113]. The γ irradiation could generate the hydroxymethyl radicals, $^{\bullet}CH_2OH$, in the reaction system, which could reduce Au(I) in Au(CN)$_2^-$ via reaction 20 and 21, and the reduced gold atoms was completely deposited on the gold particles to form larger particles. First, the radicals could transfer electrons to the gold particles and Au(CN)$_2^-$ was subsequently reduced by the stored surface electrons. Then, in the stage of particle enlargement, Au(CN)$_2^-$ was reduced in solutions with the former produced gold particles as seeds, and particles grew larger and larger up to 120 nm.

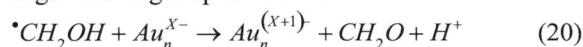

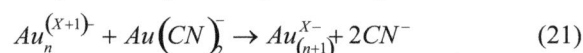

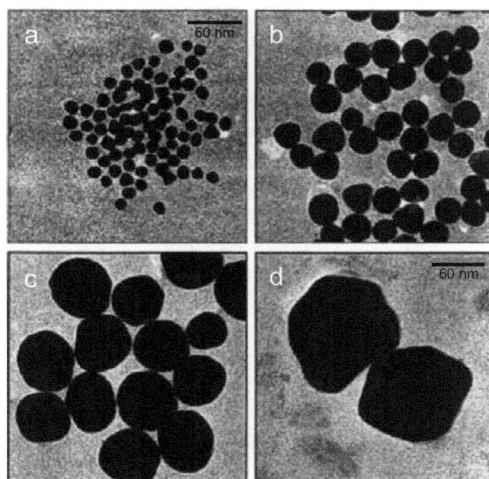

$$^{\bullet}CH_2OH + Au_n^{X-} \rightarrow Au_n^{(X+1)-} + CH_2O + H^+ \qquad (20)$$

$$Au_n^{(X+1)-} + Au\left(CN\right)_2^- \rightarrow Au_{(n+1)}^{X-} + 2CN^- \qquad (21)$$

Fig. (9). Electron micrographs showing the size-doubling of gold particles: (**a**) original citrate colloid; (**b**), (**c**) and (**d**) colloids after the first, second, and third steps of particle enlargement, respectively. Images reprinted with permission from Ref. (Henglein Langmuir 1998) [113] Copyright © 1998 The American Chemical Society.

Moreover, the conductive polyaniline-metal nanoparticles composites, consisting of polyaniline nanofibers decorated with noble-metal (Ag or Au)

nanoparticles, have been successfully synthesized through the γ radiolysis method (Pillalamarri, Blum, Tokuhiro CM 2005) [30]. After the γ rays irradiation of aqueous solution of aniline, a free-radical oxidant, and the metal salt, aniline was polymerized to form very-thin fibers with typical diameters of 50-100 nm and lengths of 1-3µm, and the metal nanoparticles formed Fig. (10). Varying the molar ratio of aniline to the metal precursor, the size and morphology of the metal nanoparticles changed from nanoparticles to micro-sized dendrites. The more the loading of metal in the composites, the higher the electrical conductivity of them. Especially, the electrical conductivity of the composites was up to 50 times higher than that of the pure polyaniline fibers.

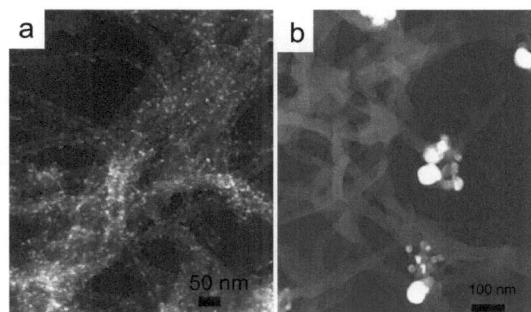

Fig. (10). (**a**) TEM image of polyaniline-Ag nanocomposites with the aniline/AgNO$_3$ mole ratio 100:1. (**b**) TEM image of polyaniline-Au composites obtained with the aniline/HAuCl$_4$ mole ratio 50:1. Images reprinted with permission from Ref. (Pillalamarri CM 2005) [30] Copyright © 2005 The American Chemical Society.

5. LASER RADIATION METHOD

A Laser radiation has been introduced to be as a preparative technique in solution chemistry, material science (Everly JACS 1978) [114] (Fojtik BBPC 1993) [115] (Yeh CL 1998) [116] (Bronstein Langmuir 1999) [117], and solid state chemistry (Tilley JMC 1999) [118]. Bard *et al.* (Everly and Traynham JACS 1978) [114] reported heterogeneous photocatalytic preparation of supported catalysts by laser photodeposition of Pt on TiO$_2$ powders and other substrates. Fendler *et al.* (Kurihara JACS 1983) [40] investigated the reduction of HAuCl$_4$ by 353-nm 60-mJ 3-5-ns laser pulse induced colloidal gold nanoparticles in water and in water-in-oil microemulsions. Henglein *et al.* (Fojtik BBPC 1993) [115] used the 694 nm light of a ruby laser to perform laser ablation of metal films such as Au and Ni, or suspended micrometer-size particles immersed in various solvents to form colloidal solutions, such as water, propanol and hexane, to produce colloidal solutions of metals. Yeh *et al.* (Yeh CL 1998) [116] reported the synthesis of Cu nanoparticles from CuO powder in 2-propanol by using a Nd:YAG laser operated at 10 Hz. Bronstein et al (Bronstein Langmuir 1999) [117] used laser photolysis for formation of Au colloids in micelle cores of block copolymer micelles derived from polystyrene-poly-vinylpyridine.

Pulsed laser irradiation method can be a novel approach to the reshaping of metal nanostructures by the inducement of the controlled heating of the nanostructures. A series of pulsed laser-induced reshaping of SiO_2-Au metallodielectric core-shell nanoparticles have been reported. (Aguirre JPCB 2004) [119] The variation of the Au nanoshells could be monitored by the transmission electron microscopy (TEM) and scanning electron microscopy (SEM) technique and the corresponding changes in the plasmon resonance frequency of the nanostructures. It was possible to control the shape of the final irradiation products via the procise adjustment of heating conditions. It was femtosecond laser irradiation, at a higher pulse intensity of 2.8 μJ and irradiation time of 30 s, the nanoshells had been destroyed, leaving only bare silica particles and gold nanospheres, and the original nanoshell plasmon band disappeared, leaving only the absorption band at 577 nm; at a lower pulse intensity of 0.28 μJ and irradiation time of 360 s, the partial reshaping of the metal shell had kept on the silica core, leaving large holes on the surface, and burning of an optical hole in the absorption spectrum. In the nanosecond laser irradiation, at high energies (0.035 mJ), nanoshells transformed into small metal colloids with a broad size distribution over a single laser shot; at the low energy limit (0.001 mJ), nanoshell destructed and produced the colloidal gold nanoparticles. Therefore, the morphology of gold nanoshells was the function of both the energy and width of the excitation pulses in the laser irradiation method.

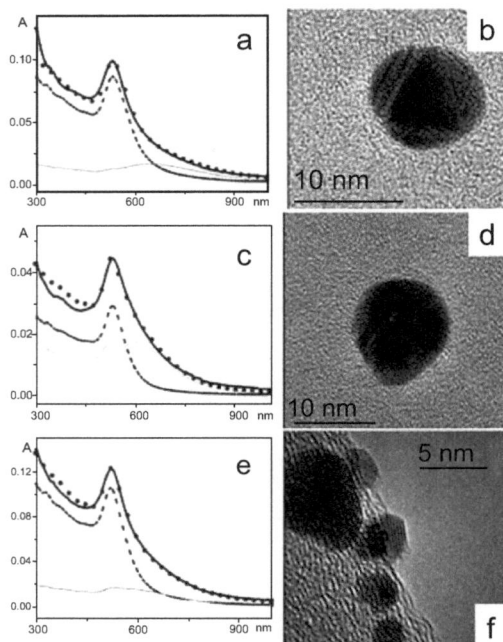

Fig. (11). (a) UV-vis spectrum of AuNP synthesized in DMSO (black line), the Mie-Gans fitting (circles), the spherical particles contribution to the fitting (dashed line) and the spheroids contribution (grey line). (b) HRTEM images of AuNP. (c) UV-vis spectrum of AuNP synthesized in THF (black line), the Mie-Gans fitting (circles), the spherical particles contribution to the fitting (dashed line)

and the spheroids contribution (grey line). (d) HRTEM images of AuNP. (e) UV-vis spectrum of AuNP synthesized in CH_3CN (black line), the Mie-Gans fitting (circles), the spherical particles contribution to the fitting (dashed line) and the spheroids contribution (grey line). (f) HRTEM images of AuNP. Images reprinted with permission from Ref. (Amendola JPCB 2006) [31] Copyright © 2006 The American Chemical Society.

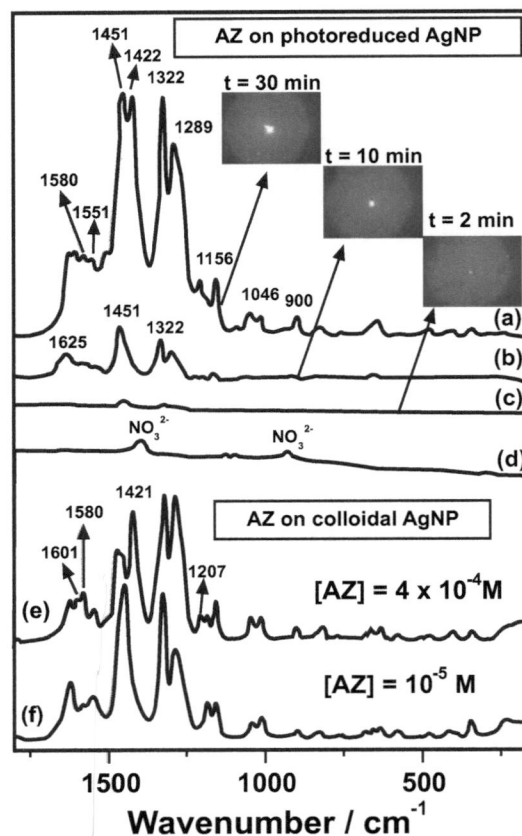

Fig. (12). SERS spectra of AZ on AgNPs prepared upon irradiation with the laser (514.5 nm, 3.0 mW) at the following irradiation times: 30 (a), 10 (b), and 2 min (c). (d) Control SERS spectrum obtained by immersion of the AgNPs obtained after 60 min irradiation (514.5 nm, 1.5 mW). (e) and (f) SERS spectra (514.5 nm)of AZ on a hydroxylamine Ag colloid at the concentrations 4×10^{-4} M (e) and 10^{-5} M (f). Images reprinted with permission from Ref. (Canamares Langmuir 2007) [120] Copyright © 2007 The American Chemical Society.

Further, a typical example is the synthesis of free and functional AuNPs by laser ablation of a gold metal plate in organic solvents, such as dimethyl sulfoxide (DMSO), acetonitrile, and tetrahydrofuran (THF). (Amendola JPCB 2006) [31] UV-vis spectroscopy and TEM could characterize the nanoparticles. While the laser ablation occurred in the DMSO, THF, and CH_3CN, the surface plasmon absorption of AuNPs was centered at 530 nm, 528 nm, and 522 nm, respectively, and the average radius of AuNPs was 2.4±0.9 nm, 4.1±2.5 nm, and 1.8±1.2 nm, respectively Fig. (11). Importantly, the theoretical modes of the Mie model for spherical particles and the Gans model for spheroids could effectively interpret the experimental UV-vis spectra of AuNPs.

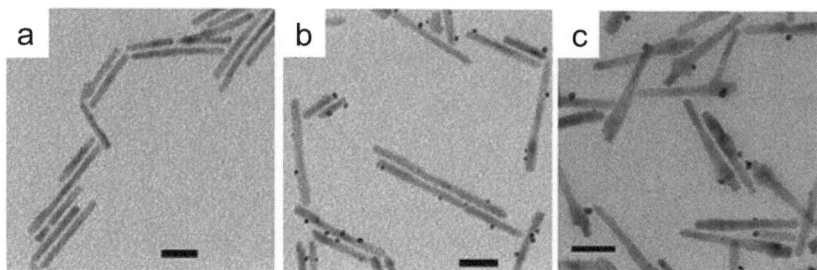

Fig. (13). TEM images of (**a**) CdS nanorods before irradiation, (**b**) after photodeposition. After exposure to light, nanoparticles appear along the length of the nanorods. Scale bar = 20 nm. (**c**) the resulting heterostructures after photodeposition of Pt on CdSe/CdS core/shell nanorods. Scale bar = 20 nm. Images reprinted with permission from Ref. (Dukovic AM 2008) [104] Copyright © 2008 Wiley-VCH.

Interestingly, Ag nanoparticles (AgNPs) have been prepared by laser photoreduction as substrates for in situ surface-enhanced Raman scattering (SERS) analysis of dyes Fig. (**12**). (Canamares Langmuir 2007) [120] AgNPs were prepared and immobilized on a water/solid interface where the aqueous phase contained the Ag^+ cations and the solid surface was of hydrophilic nature (glass and cellulose). The size of AgNPs increased with the rise of the irradiation time, the laser power, and the metallic cation concentration. The morphology of AgNPs mainly included the spherical and planar shape, and the planar shape were formed after 15 min of laser irradiation and were responsible for the remarkable SERS intensification. The prepared AgNPs demonstrated a high SERS effectiveness in the detection of dispersed adsorbates such as the anthraquinonic dye alizarin. Importantly, while the dye had placed on a hydrophilic substrate, the photoreduced AgNPs could be employed for in situ detection of them.

Laser irradiation can induce the photodeposition of metal on semiconductor nanostructures for the formation of metal-semiconductor nanocomposites (Dukovic AM 2008) [104]. The reaction mechanism is similar with the UV photodeposition process. The laser irradiation of the solution containing semiconductor nanostructures can produce the conduction band electrons and the valence band holes, which can reduce metal ions and oxide the hole scavengers. The semiconductor band structure, spatial organization and surface chemistry are important factors in the photodeposition process.

Alivisatos *et al.* (Dukovic AM 2008) [104] have reported the photodeposition of Pt on colloidally synthesized CdS and CdSe/CdS nanorods by the laser irradiation Fig. (**13**). The deposition of Pt on each CdS nanorod with the number of Pt nanoparticles ranging from 0 to 6 was highly hererogeneous. The average diameter of the Pt nanoparticels ranged from 1.5 to 2.7 nm. In contrast, most CdSe/CdS core/shell nanorods located only one metal nanoparticle near the CdSe core. The octylamine-treated CdS nanorods couldn't deposit Pt nanoparticles. The yield of Pt deposited on CdS nanorods was dependent on the irradiation power and reactant concentrations. This method should provide insights useful for the development of photochemistry of metal-semiconductor nanocomposites into a synthetic tool.

6. CONCLUSIONS

In conclusion, much progress has been made by photoreduction technique for preparation of metal nanostructures. γ rays or UV radiation method has extensively exploited for synthesis of various nanostructured materials such as metal-inorganic nanocomposites, and metal-polymer nanocomposites. In contrast, the reports on the fabrication of nanostructured materials by the laser radiation technique in solution systems are few and most only focused on metal colloidal particles or alloys.

Exploration of both sizes and shape controlled methods is still both challenging and interesting fields in future. From the point of view of application, how to integrate the photoreduction technique with solution processing is worth exploring. Especially, how to initiate some novel inorganic reactions on the interface of solid substrate and liquid in solution processing system at mild conditions, i. e., even at room temperature, is the key point. We believe that some reactions may be initiated and activated by photo irradiation in solution system for the synthesis of metal nanostructures. Indeed, recent research has demonstrated that metal nanostructures could be accurately synthesized by the photoreduction method, which is an ongoing area of research that requires strong interdisciplinary collaboration of researchers from different fields.

7. ACKNOWLEDGEMENTS

This work was supported by the national science foundation of China (NSFC, 50732006, 20621061, 20671085), 2005CB623601, and the Partner Group of the CAS-MPG.

8. REFERENCES

[1]　　Ajayan PM, Iijima S. Capillarity-Induced Filling of Carbon Nanotubes. Nature 1993; 361(6410): 333-4.

[2]　　Ahmadi TS, Wang ZL, Green TC, Henglein A, ElSayed MA. Shape-controlled synthesis of colloidal platinum nanoparticles. Science 1996; 272(5270): 1924-6.

[3]   Andres RP, Bielefeld JD, Henderson JI, Janes DB, Kolagunta VR, Kubiak CP, Mahoney WJ,Osifchin RG. Self-assembly of a two-dimensional superlattice of molecularly linked metal clusters. Science 1996; 273(5282): 1690-3.

[4]   Antonietti M,Goltner C. Superstructures of functional colloids: Chemistry on the nanometer scale. Angew Chem Int Edit 1997; 36(9): 910-28.

[5]   Antonietti M, Grohn F, Hartmann J,Bronstein L. Nonclassical shapes of noble-metal colloids by synthesis in microgel nanoreactors. Angew Chem Int Edit 1997; 36(19): 2080-3.

[6]   Guo SH, Tsai SJ, Kan HC, Tsai DH, Zachariah MR,Phaneuf RJ. The effect of an active substrate on nanoparticle-enhanced fluorescence. Adv Mater 2008; 20(8): 1424-8.

[7]   Nehl CL,Hafner JH. Shape-dependent plasmon resonances of gold nanoparticles. J Mater Chem 2008; 18(21): 2415-9.

[8]   Tao AR, Habas S,Yang PD. Shape control of colloidal metal nanocrystals. Small 2008; 4(3): 310-25.

[9]   Teng XW, Han WQ, Wang Q, Li L, Frenkel AI,Yang JC. Hybrid Pt/Au nanowires: Synthesis and electronic structure. J Phys Chem C 2008; 112(38): 14696-701.

[10]   Nishida N, Yao H,Kimura K. Chiral functionalization of optically inactive monolayer-protected silver nanoclusters by chiral ligand-exchange reactions. Langmuir 2008; 24(6): 2759-66.

[11]   Didiot C, Pons S, Kierren B, Fagot-Revurat Y,Malterre D. Nanopatterning the electronic properties of gold surfaces with self-organized superlattices of metallic nanostructures. Nat Nanotechnol 2007; 2(10): 617-21.

[12]   Moffitt M,Eisenberg A. Size Control of Nanoparticles in Semiconductor-Polymer Composites .1. Control Via Multiplet Aggregation Numbers in Styrene-Based Random Ionomers. Chem Mater 1995; 7(6): 1178-84.

[13]   Moffitt M, Mcmahon L, Pessel V,Eisenberg A. Size Control of Nanoparticles in Semiconductor-Polymer Composites .2. Control Via Sizes of Spherical Ionic Microdomains in Styrene-Based Diblock Ionomers. Chem Mater 1995; 7(6): 1185-92.

[14]   Foss CA, Tierney MJ,Martin CR. Template Synthesis of Infrared-Transparent Metal Microcylinders - Comparison of Optical-Properties with the Predictions of Effective Medium Theory. J Phys Chem-Us 1992; 96(22): 9001-7.

[15]   Nakao Y. Noble-Metal Solid Sols in Poly(Methyl Methacrylate). J Colloid Interf Sci 1995; 171(2): 386-91.

[16]   Zhu YJ, Qian YT, Li XJ,Zhang MW. gamma-Radiation synthesis and characterization of polyacrylamide-silver nanocomposites. Chem Commun 1997; 12: 1081-2.

[17]   Yin YD, Xu XL, Xia CJ, Ge XW, Zhang ZC. Synthesis and characterization of poly(butyl acrylate-co-styrene)-silver nanocomposites by gamma radiation in W/O microemulsions. Chem Commun 1998; 8: 941-2.

[18]   Wang Y, Liu HF,Jiang YY. A New Method for Immobilization of Polymer-Protective Colloidal Platinum Metals Via Co-Ordination Capture with Anchored Ligands - Synthesis of the 1st Example of a Mercapto-Containing Supported Metallic Catalyst for Hydrogenation of Alkenes with High-Activity. J Chem Soc Chem Comm 1989; 24: 1878-9.

[19]   Ohtaki M, Komiyama M, Hirai H, Toshima N. Effects of polymer support on the substrate selectivity of covalently immobilized ultrafine rhodium particles as a catalyst for olefin hydrogenation. Macromolecules 1991; 24(20): 5567-72.

[20]   Toshima N,Wang Y. Preparation and Catalysis of Novel Colloidal Dispersions of Copper/Noble Metal Bimetallic Clusters, Langmuir 1994; 10: 4574-80

[21]   Chen CW, Chen MQ, Serizawa T, Akashi M. In situ synthesis and the catalytic properties of platinum colloids on polystyrene microspheres with surface-grafted poly(N-isopropylacrylamide). Chem Commun 1998; 7: 831-2.

[22]   Chen CW, Chen MQ, Serizawa T,Akashi M. In-situ formation of silver nanoparticles on poly(N-isopropylacrylamide)-coated polystyrene microspheres. Adv Mater 1998; 10(14): 1122-6.

[23]   Sarathy KV, Kulkarni GU, Rao CNR. A novel method of preparing thiol-derivatised nanoparticles of gold, platinum and silver forming superstructures. Chem Commun 1997; 6: 537-8.

[24]   Selvan ST. Novel nanostructures of gold-polypyrrole composites. Chem Commun 1998; 3: 351-2.

[25]   Selvan ST, Spatz JP, Klok HA, Moller M. Gold-polypyrrole core-shell particles in diblock copolymer micelles. Adv Mater 1998; 10(2): 132-4.

[26]   Shenhar R, Norsten TB,Rotello VM. Polymer-mediated nanoparticle assembly: Structural control and applications. Adv Mater 2005; 17(6): 657-69.

[27]   Tang ZY,Kotov NA. One-dimensional assemblies of nanoparticles: Preparation, properties, and promise. Adv Mater 2005; 17(8): 951-62.

[28]   Yang SM, Kim SH, Lim JM,Yi GR. Synthesis and assembly of structured colloidal particles. J Mater Chem 2008; 18(19): 2177-90.

[29]   Yu SH, Colfen H,Mastai Y. Formation and optical properties of gold nanoparticles synthesized in the presence of double-hydrophilic block copolymers. J. Nanosci. Nanotechnol. 2004; 4(3): 291-8.

[30]   Pillalamarri SK, Blum FD, Tokuhiro AT,Bertino MF. One-pot synthesis of polyaniline - Metal nanocomposites. Chem Mater 2005; 17(24): 5941-4.

[31]   Amendola V, Polizzi S,Meneghetti M. Laser ablation synthesis of gold nanoparticles in organic solvents. J Phys Chem B 2006; 110(14): 7232-7.

[32]   Zhang B, Hou WY, Ye XC, Fu SQ,Xie Y. 1D tellurium nanostructures: Photothermally assisted morphology-controlled synthesis and applications in preparing functional nanoscale materials. Adv Funct Mater 2007; 17(3): 486-92.

[33]   Esumi K, Matsuhisa K,Torigoe K. Preparation of Rodlike Gold Particles by Uv Irradiation Using Cationic Micelles as a Template. Langmuir 1995; 11(9): 3285-7.

[34]   Chang S-S, Shih C-W, Chen C-D, Lai W-C,Wang CRC. The Shape Transition of Gold Nanorods, 1999; 15: 701-9

[35]   Curtis AC, Duff DG, Edwards PP, Jefferson DA, Johnson BFG, Kirkland AI,Wallace A. A Morphology-Selective Copper Organosol, 1988; 27: 1530-3

[36]   Mayer A,Antonietti M. Investigation of polymer-protected noble metal nanoparticles by transmission electron microscopy: control of particle morphology and shape. Colloid Polym Sci 1998; 276(9): 769-79.

[37]   Milligan WO,Morriss RH. Morphology of Colloidal Gold--A Comparative Study, 1964; 86: 3461-7

[38]   Sarathy KV, Raina G, Yadav RT, Kulkarni GU, Rao CNR. Thiol-derivatized nanocrystalline arrays of gold, silver, and platinum. J Phys Chem B 1997; 101(48): 9876-80.

[39]   Mosseri S, Henglein A,Janata E. Reduction of dicyanoaurate(I) in aqueous solution: formation of nonmetallic clusters and colloidal gold, 1989; 93: 6791-5

[40]   Kurihara K, Kizling J, Stenius P, Fendler JH. Laser and pulse radiolytically induced colloidal gold formation in water and in water-in-oil microemulsions, 1983; 105: 2574-9

[41]   Baxendale JH,Busi F. The Study of Fast Processes and Transient Species by Electron Pulse Radiolysis. NATO ASI, D. Reidel T, Series 86, 1982.

[42]   Belloni J, Mostafavi M, Remita H, Marignier JL,Delcourt MO. Radiation-induced synthesis of mono- and multi-metallic clusters and nanocolloids. New J Chem 1998; 22(11): 1239-55.

[43]   Henglein A. Physicochemical Properties of Small Metal Particles in Solution - Microelectrode Reactions, Chemisorption, Composite Metal Particles, and the Atom-to-Metal Transition. J Phys Chem-Us 1993; 97(21): 5457-71.

[44]   Itakura T, Torigoe K,Esumi K. Preparation and Characterization of Ultrafine Metal Particles in Ethanol by Uv Irradiation Using a Photoinitiator. Langmuir 1995; 11(10): 4129-34.

[45]   Yonezawa Y, Sato T, Kuroda S,Kuge K. Photochemical Formation of Colloidal Silver - Peptizing Action of

Acetone Ketyl Radical. J Chem Soc Faraday T 1991; 87(12): 1905-10.

[46]   Zhou Y, Yu SH, Wang CY, Li XG, Zhu YR,Chen ZY. A novel ultraviolet irradiation photoreduction technique for the preparation of single-crystal Ag nanorods and Ag dendrites. Adv Mater 1999; 11(10): 850-2.

[47]   Zhou Y, Wang CY, Zhu YR, Chen ZY. A novel ultraviolet irradiation technique for shape-controlled synthesis of gold nanoparticles at room temperature. Chem Mater 1999; 11(9): 2310-2.

[48]   Harada M, Takahashi S. Synthesis of ruthenium particles by photoreduction in polymer solutions. J Colloid Interf Sci 2008; 325(1): 1-6.

[49]   Tan S, Erol M, Attygalle A, Du H, Sukhishvili S. Synthesis of positively charged silver nanoparticles via photoreduction of AgNO3 in branched polyethyleneimine/HEPES solutions. Langmuir 2007; 23(19): 9836-43.

[50]   Tamai T, Watanabe M, Hatanaka Y, Tsujiwaki H, Nishioka N,Matsukawa K. Formation of Metal Nanoparticles on the Surface of Polymer Particles Incorporating Polysilane by UV Irradiation. Langmuir 2008; 24(24): 14203-8.

[51]   Wang TC, Rubner MF,Cohen RE. Polyelectrolyte multilayer nanoreactors for preparing silver nanoparticle composites: Controlling metal concentration and nanoparticle size. Langmuir 2002; 18(8): 3370-5.

[52]   Joly S, Kane R, Radzilowski L, Wang T, Wu A, Cohen RE, Thomas EL,Rubner MF. Multilayer nanoreactors for metallic and semiconducting particles. Langmuir 2000; 16(3): 1354-9.

[53]   Toshima N, Takahashi T. Colloidal Dispersions of Platinum and Palladium Clusters Embedded in the Micelles - Preparation and Application to the Catalysis for Hydrogenation of Olefins. B Chem Soc Jpn 1992; 65(2): 400-9.

[54]   Sato T, Kuroda S, Takami A, Yonezawa Y,Hada H. Photochemical Formation of Silver Gold (Ag-Au) Composite Colloids in Solutions Containing Sodium Alginate. Appl Organomet Chem 1991; 5(4): 261-8.

[55]   Torigoe K, Esumi K. Preparation of Bimetallic Ag-Pd Colloids from Silver(I) Bis(Oxalato)Palladate(Ii). Langmuir 1993; 9(7): 1664-7.

[56]   Torigoe K,Esumi K. Preparation of Colloidal Gold by Photoreduction of Aucl4--Cationic Surfactant Complexes. Langmuir 1992; 8(1): 59-63.

[57]   Torigoe K,Esumi K. Formation of Nonspherical Palladium Nanocrystals in Sds Poly(Acrylamide) Gel. Langmuir 1995; 11(11): 4199-201.

[58]   Mayer A,Antonietti M. Polymer-protected noble metal nanoparticles: Control of particle morphology and shape. Abstr Pap Am Chem S 1998; 216(U632-U.

[59]   Sakamoto Y, Fukuoka A, Higuchi T, Shimomura N, Inagaki S,Ichikawa M. Synthesis of platinum nanowires in organic-inorganic mesoporous silica templates by photoreduction: Formation mechanism and isolation. J Phys Chem B 2004; 108(3): 853-8.

[60]   Fukuoka A, Higashimoto N, Sakamoto Y, Inagaki S, Fukushima Y,Ichikawa M. Preparation and catalysis of Pt and Rh nanowires and particles in FSM-16. Micropor Mesopor Mat 2001; 48(1-3): 171-9.

[61]   Yu JS, Kim JY, Lee S, Mbindyo JKN, Martin BR,Mallouk TE. Template synthesis of polymer-insulated colloidal gold nanowires with reactive ends. Chem Commun 2000; 24: 2445-6.

[62]   Foss CA, Hornyak GL, Stockert JA, Martin CR. Optical-Properties of Composite Membranes Containing Arrays of Nanoscopic Gold Cylinders. J Phys Chem-Us 1992; 96(19): 7497-9.

[63]   Grobert N, Mayne M, Terrones M, Sloan J, Dunin-Borkowski RE, Kamalakaran R, Seeger T, Terrones H, Ruhle M, Walton DRM, Kroto HW,Hutchison JL. Alloy nanowires: Invar inside carbon nanotubes. Chem Commun 2001; 5: 471-2.

[64]   Govindaraj A, Satishkumar BC, Nath M,Rao CNR. Metal nanowires and intercalated metal layers in single-walled carbon nanotube bundles. Chem Mater 2000; 12(1): 202-5.

[65]   Sloan J, Hammer J, Zwiefka-Sibley M,Green MLH. The opening and filling of single walled carbon nanotubes (SWTs). Chem Commun 19983): 347-8.

[66]   Yanagisawa T, Shimizu T, Kuroda K,Kato C. The Preparation of Alkyltrimethylammonium-Kanemite Complexes and Their Conversion to Microporous Materials. B Chem Soc Jpn 1990; 63(4): 988-92.

[67]   Inagaki S, Fukushima Y,Kuroda K. Synthesis of Highly Ordered Mesoporous Materials from a Layered Polysilicate. J Chem Soc Chem Comm 19938): 680-2.

[68]   Kresge CT, Leonowicz ME, Roth WJ, Vartuli JC,Beck JS. Ordered Mesoporous Molecular-Sieves Synthesized by a Liquid-Crystal Template Mechanism. Nature 1992; 359(6397): 710-2.

[69]   Zhao DY, Feng JL, Huo QS, Melosh N, Fredrickson GH, Chmelka BF,Stucky GD. Triblock copolymer syntheses of mesoporous silica with periodic 50 to 300 angstrom pores. Science 1998; 279(5350): 548-52.

[70]   Fukuoka A, Araki H, Sakamoto Y, Inagaki S, Fukushima Y,Ichikawa M. Palladium nanowires and nanoparticles in mesoporous silica templates. Inorg Chim Acta 2003; 350: 371-8.

[71]   Fukuoka A, Sakamoto Y, Guan S, Inagaki S, Sugimoto N, Fukushima Y, Hirahara K, Iijima S,Ichikawa M. Novel templating synthesis of necklace-shaped mono- and bimetallic nanowires in hybrid organic-inorganic mesoporous material. J Am Chem Soc 2001; 123(14): 3373-4.

[72]   Banerjee IA, Yu LT,Matsui H. Cu nanocrystal growth on peptide nanotubes by biomineralization: Size control of Cu nanocrystals by tuning peptide conformation. P Natl Acad Sci USA 2003; 100(25): 14678-82.

[73]   Sun LL, Wei G, Song YH, Liu ZG, Wang L,Li ZA. Fabrication of silver nanoparticles ring templated by plasmid DNA. Appl Surf Sci 2006; 252(14): 4969-74.

[74]   Aldaye FA, Palmer AL,Sleiman HF. Assembling materials with DNA as the guide. Science 2008; 321(5897): 1795-9.

[75]   Flynn CE, Lee SW, Peelle BR,Belcher AM. Viruses as vehicles for growth, organization and assembly of materials. Acta Mater 2003; 51(19): 5867-80.

[76]   Antonietti M,Ozin GA. Promises and problems of mesoscale materials chemistry or why meso? Chem-Eur J 2004; 10(1): 29-41.

[77]   Zhong ZY, Sim DH, Teo J, Luo JZ, Zhang HJ,Gedanken A. D-glucose-derived polymer intermediates as templates for the synthesis of ultrastable and redispersible gold colloids. Langmuir 2008; 24(9): 4655-60.

[78]   Samson J, Varotto A, Nahirney PC, Toschi A, Piscopo I,Drain CM. Fabrication of Metal Nanoparticles Using Toroidal Plasmid DNA as a Sacrificial Mold. ACS Nano 2009; 3(2): 339-44.

[79]   Link S, Wang ZL,El-Sayed MA. Alloy formation of gold-silver nanoparticles and the dependence of the plasmon absorption on their composition. J Phys Chem B 1999; 103(18): 3529-33.

[80]   Treguer M, de Cointet C, Remita H, Khatouri J, Mostafavi M, Amblard J, Belloni J,de Keyzer R. Dose rate effects on radiolytic synthesis of gold-silver bimetallic clusters in solution. J Phys Chem B 1998; 102(22): 4310-21.

[81]   deCointet C, Mostafavi M, Khatouri J,Belloni J. Growth and reactivity of silver clusters in cyanide solution. J Phys Chem B 1997; 101(18): 3512-6.

[82]   Schmid G, Lehnert A, Malm JO,Bovin JO. Ligand-Stabilized Bimetallic Colloids Identified by Hrtem and Edx. Angew Chem Int Edit 1991; 30(7): 874-6.

[83]   Mallik K, Mandal M, Pradhan N,Pal T. Seed mediated formation of bimetallic nanoparticles by UV irradiation: A photochemical approach for the preparation of "core-shell" type structures. Nano Lett 2001; 1(6): 319-22.

[84]   Mandal S, Selvakannan PR, Pasricha R,Sastry M. Keggin ions as UV-switchable reducing agents in the synthesis of Au core-Ag shell nanoparticles. J Am Chem Soc 2003; 125(28): 8440-1.

[85]   Chen CW, Serizawa T,Akashi M. Preparation of platinum colloids on polystyrene nanospheres and their catalytic

properties in hydrogenation. Chem Mater 1999; 11(5): 1381-9.

[86]   Dokoutchaev A, James JT, Koene SC, Pathak S, Prakash GKS,Thompson ME. Colloidal metal deposition onto functionalized polystyrene microspheres. Chem Mater 1999; 11(9): 2389-99.

[87]   Kim F, Song JH,Yang PD. Photochemical synthesis of gold nanorods. J Am Chem Soc 2002; 124(48): 14316-7.

[88]   Korchev AS, Konovalova T, Cammarata V, Kispert L, Slaten L,Mills G. Radical-induced generation of small silver particles in SPEEK/PVA polymer films and solutions: UV-vis, EPR, and FT-IR studies. Langmuir 2006; 22(1): 375-84.

[89]   Li Y, Boone E,El-Sayed MA. Size effects of PVP-Pd nanoparticles on the catalytic Suzuki reactions in aqueous solution. Langmuir 2002; 18(12): 4921-5.

[90]   Lu Y, Mei Y, Drechsler M,Ballauff M. Thermosensitive core-shell particles as carriers for Ag nanoparticles: Modulating the catalytic activity by a phase transition in networks. Angew. Chem.-Int. Edit. 2006; 45(5): 813-6.

[91]   Mayer ABR, Grebner W,Wannemacher R. Preparation of silver-latex composites. J Phys Chem B 2000; 104(31): 7278-85.

[92]   Mayer ABR,Mark JE. Immobilization of palladium nanoparticles on latex supports and their potential for catalytic applications. Angew. Makromol. Chem. 1999; 268:52-8.

[93]   Mei Y, Lu Y, Polzer F, Ballauff M,Drechsler M. Catalytic activity of palladium nanoparticles encapsulated in spherical polyelectrolyte brushes and core-shell microgels. Chem Mater 2007; 19(5): 1062-9.

[94]   Mei Y, Sharma G, Lu Y, Ballauff M, Drechsler M, Irrgang T,Kempe R. High catalytic activity of platinum nanoparticles immobilized on spherical polyelectrolyte brushes. Langmuir 2005; 21(26): 12229-34.

[95]   Pathak S, Greci MT, Kwong RC, Mercado K, Prakash GKS, Olah GA,Thompson ME. Synthesis and applications of palladium-coated poly(vinylpyridine) nanospheres. Chem Mater 2000; 12(7): 1985-9.

[96]   Sakamoto M, Tachikawa T, Fujitsuka M,Majima T. Three-dimensional writing of copper nanoparticles in a polymer matrix with two-color laser beams. Chem Mater 2008; 20(6): 2060-2.

[97]   Schuetz P,Caruso F. Semiconductor and metal nanoparticle formation on polymer spheres coated with weak polyelectrolyte multilayers. Chem Mater 2004; 16(16): 3066-73.

[98]   Shi WL, Sahoo Y, Swihart MT,Prasad PN. Gold nanoshells on polystyrene cores for control of surface plasmon resonance. Langmuir 2005; 21(4): 1610-7.

[99]   Suzuki D,Kawaguchi H. Modification of gold nanoparticle composite nanostructures using thermosensitive core-shell particles as a template. Langmuir 2005; 21(18): 8175-9.

[100] Wen F, Zhang WQ, Wei GW, Wang Y, Zhang JZ, Zhang MC,Shi LQ. Synthesis of noble metal nanoparticles embedded in the shell layer of core-shell poly(styrene-co-4-vinylpyridine) micospheres and their application in catalysis. Chem Mater 2008; 20(6): 2144-50.

[101] Bruchez M, Moronne M, Gin P, Weiss S,Alivisatos AP. Semiconductor nanocrystals as fluorescent biological labels. Science 1998; 281(5385): 2013-6.

[102] Huynh WU, Dittmer JJ,Alivisatos AP. Hybrid nanorod-polymer solar cells. Science 2002; 295(5564): 2425-7.

[103] Pacholski C, Kornowski A,Weller H. Site-specific photodeposition of silver on ZnO nanorods. Angew. Chem.-Int. Edit. 2004; 43(36): 4774-7.

[104] Dukovic G, Merkle MG, Nelson JH, Hughes SM,Alivisatos AP. Photodeposition of Pt on Colloidal CdS and CdSe/CdS Semiconductor Nanostructures. Adv Mater 2008; 20(22): 4306-11.

[105] Cozzoli PD, Comparelli R, Fanizza E, Curri ML, Agostiano A,Laub D. Photocatalytic Synthesis of Silver Nanoparticles Stabilized by TiO2 Nanorods: A Semiconductor/Metal Nanocomposite in Homogeneous Nonpolar Solution. J Am Chem Soc 2004; 126(12): 3868- 79.

[106] Ohtani B, Ogawa Y,Nishimoto S. Photocatalytic activity of amorphous-anatase mixture of titanium(IV) oxide particles suspended in aqueous solutions. J Phys Chem B 1997; 101(19): 3746-52.

[107] Ng YH, Ikeda S, Harada T, Higashida S, Sakata T, Mori H,Matsumura M. Fabrication of hollow carbon nanospheres encapsulating platinum nanoparticles using a photocatalytic reaction. Adv Mater 2007; 19(4): 597-601.

[108] Marignier JL, Belloni J, Delcourt MO,Chevalier JP. Microaggregates of non-noble metals and bimetallic alloys prepared by radiation-induced reduction. Nature 1985; 317(6035): 344-5.

[109] Zhu YJ, Qian YT, Zhang MW, Chen ZY, Lu B,Wang CS. Preparation of Nanocrystalline Silver Powders by Gamma-Ray Radiation Combined with Hydrothermal Treatment. Mater Lett 1993; 17(5): 314-8.

[110] Zhu YJ, Qian YT, Zhang MW, Chen ZY,Zhou GE. Gamma-Radiation Sol-Gel Synthesis of Glass-Metal Nanocomposites. J Mater Chem 1994; 4(10): 1619-20.

[111] Zhu YG, Qian YT, Zhang MW, Chen ZY, Lu B,Zhou G. Gamma-Radiation Hydrothermal Synthesis and Characterization of Nanocrystalline Copper Powders. Mat Sci Eng B-Solid 1994; 23(2): 116-9.

[112] Zhu YJ, Qian YT, Huang H, Zhang MW,Liu SX. Sol-gel gamma-radiation synthesis of titania-silver nanocomposites. Mater Lett 1996; 28(4-6): 259-61.

[113] Henglein A,Meisel D. Radiolytic control of the size of colloidal gold nanoparticles. Langmuir 1998; 14(26): 7392-6.

[114] Everly CR,Traynham JG. Formation and rearrangement of ipso intermediates in aromatic free-radical chlorination reactions. J Am Chem Soc 1978; 100(13): 4316-7.

[115] Fojtik A,Henglein A. Laser Ablation of Films and Suspended Particles in a Solvent - Formation of Cluster and Colloid Solutions. Ber Bunsen Phys Chem 1993; 97(2): 252-4.

[116] Yeh YH, Yeh MS, Lee YP,Yeh CS. Formation of Cu nanoparticles from CuO powder by laser ablation in 2-propanol. Chem Lett 1998; 11: 1183-4.

[117] Bronstein L, Chernyshov D, Valetsky P, Tkachenko N, Lemmetyinen H, Hartmann J,Forster S. Laser Photolysis Formation of Gold Colloids in Block Copolymer Micelles. Langmuir 1999; 15(1): 83-91.

[118] Tilley RJD. Ultra-rapid laser irradiation as a preparative technique in solid state chemistry. J Mater Chem 1999; 9(1): 259-63.

[119] Aguirre CM, Moran CE, Young JF,Halas NJ. Laser-induced reshaping of metallodielectric nanoshells under femtosecond and nanosecond plasmon resonant illumination. J Phys Chem B 2004; 108(22): 7040-5.

[120] Canamares MV, Garcia-Ramos JV, Gomez-Varga JD, Domingo C,Sanchez-Cortes S. Ag nanoparticles prepared by laser photoreduction as substrates for in situ surface-enhanced raman scattering analysis of dyes. Langmuir 2007; 23(9): 5210-5.

CHAPTER 5

SELF-ASSEMBLY OF NANOSTRUCTURES

Mustafa Çulha

Yeditepe University, Faculty of Engineering and Architecture Genetics and Bioengineering Department, Kayisdagi, Istanbul, Turkey; Email: mculha@yeditepe.edu.tr

Abstract: This chapter is mainly focused on current novel approaches that have been made to synthesize metal nanoparticles with predetermined shape, size, and fair stability using self assembly process. It has also demonstrated the influence of various inorganic, organic compounds, polymer, and biological molecules on the nanostructures of particles.

Key words: Self-assembly, nano-assembly, nanostructure, nanoparticles, agglomeration.

1. INTRODUCTION

The extraordinary physicochemical properties of nanometer size particles due to quantum size effect and their astonishing collective properties upon their assembly into organized structures and patterns are the main reasons behind the current research efforts for the preparation of new nanoparticles from diverse materials [1-5]. The size and shape dependence of physicochemical properties provide additional opportunities [5-10] for their applications in diverse fields such as sensing [11], diagnostics [12], nanoelectronics [13], and catalysis [14]. The two fundamental problems associated with the nanotechnology concept are the preparation uniform size and shape of nanoparticles and their precise assembly into desired 1D, 2D or 3D structures for novel applications. The first problem has been partially resolved and a number of uniform nanoparticles synthesis and preparation methods are recently reported [15-22]. However, the latter problem remains to be a genuine challenge while the importance of precise and defect free assembly of nanoparticles into different structures or patterns is indisputable due to the fact that the physicochemical properties show unusual properties with their organization. Therefore, the self-assembly concept of nanostructures is under intense investigation. As the human beings have been using the "top-down" approach since their existence, the "bottom-up" approach is in their infancy yet and their potentials need to be fully explored for its promise. Although this chapter mostly focuses on the "bottom-up" for the self-assembly of the nanostructures, the "top-down" approach is also included in some cases where the two approaches are combined. In each section, the recent advances and applications in the self-assembly concept are discussed following a brief introduction of the fundamentals.

The preparation of nanostructures and patterns on surfaces can be achieved using several conventional and unconventional techniques such as lithography [23] and microcontact-printing [24], which are considered as the "top-down" approaches. However, with these advanced techniques, the preparation of long-range and defect-free areas is currently cumbersome or expensive. The self-assembly for the organization of nanoparticles into 1D, 2D and 3D as a bottom-up approach recently emerges.

It should be noted that the surface properties of nanoparticles play a key role during their self-assembly into the target structures [25-30]. Therefore, it is necessary to pay special attention to the surface chemistry of the nanostructures, which determines the final properties in solutions and at interfaces through the weak non-covalent interactions such as hydrogen bonding van der Waals interactions, dipole-dipole interactions. The surface property of the nanoparticles is determined by the synthesis method or it can be altered with chemical attachment or physical adsorption of the desired molecules. A diverse number of methods for synthesis and surface chemistry are available for derivatization [15-22].

2. SELF-ASSEMBLY AT SOLID-LIQUID INTERFACE

The majority of nanoparticles are synthesized and their assembly is achieved from their solutions or suspensions. Several approaches have been explored to assemble the nanoparticles into 1D, 2D and 3D structures on solid surfaces, which constitute the solid-liquid interface. The self-assembly is achieved during the evaporation of solvent from a droplet located on a surface. Not only the physically and chemically prepared templates but also spontaneous assembly using the existing week forces in solution can be utilized to assemble the nanoparticles onto surfaces. In this section, the recent self-assembly reports are discussed in three different sections; assembly from a drying droplet, template assisted assembly, and programmed self-assembly.

3. SELF-ASSEMBLY FROM A DRYING DROPLET

The basis of the assembly is the weak force that becomes dominant among the nanoparticles as the

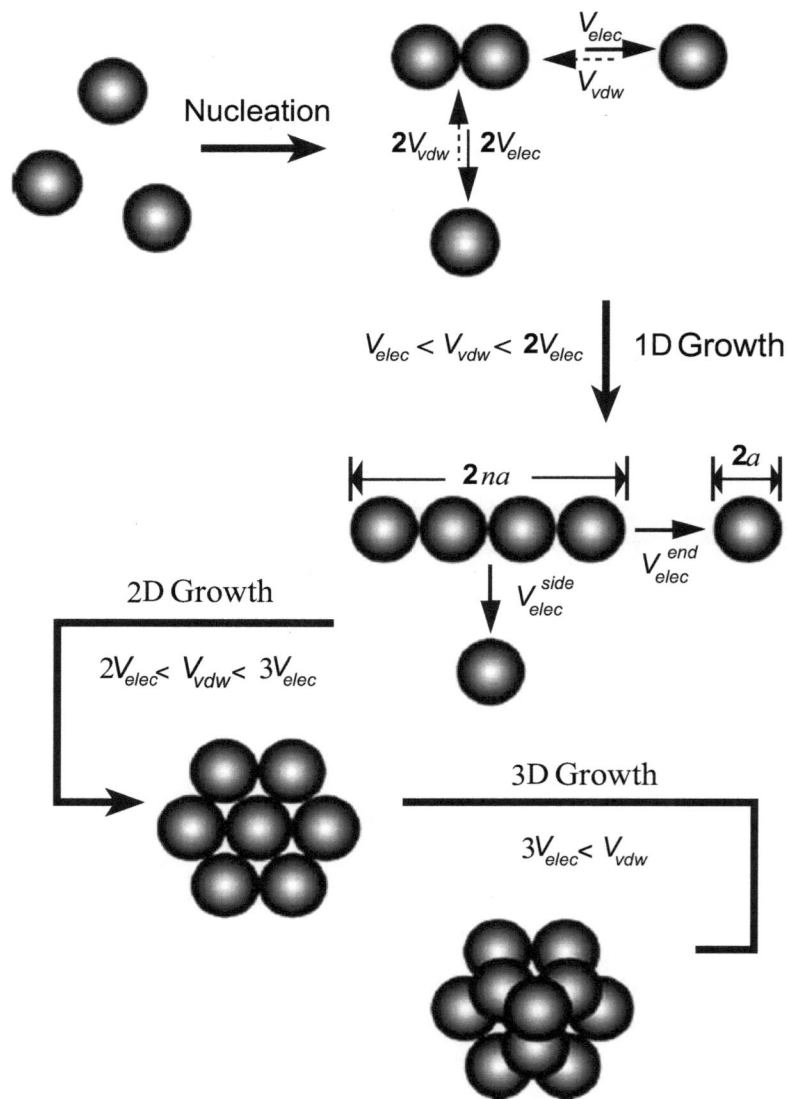

Fig. (1). The demonstration of the growth of the 1D or 2D structures from gold nanoparticles under the influence of isotropic long-range electrostatic repulsion and isotropic short-range van der Waals attractions in the presence of short-range anisotropic dipolar attraction forces. Reprinted with permission from Ref. 38. Copyright @Wiley-VCH Verlag GmbH & Co. KGaA.

solvent is evaporated from a droplet of suspension. In a stabilized colloidal suspension, the nanoparticles remain well dispersed. As the solvent from the top of the suspension starts to evaporate, the nanoparticles concentration increases at the immediate below the liquid surface due to the faster the evaporation rate than the diffusion of nanoparticles causing the nanoparticles to assemble. This phenomenon was first reported by Denkov *et al.* and the driving mechanism of the formation of irreversible packing nanoparticles was explained by the thermodynamics of non-equilibrium processes of aggregating nanoparticles [32, 33]. The 2D and 3D structures with the use of gold, silver and polystyrene particles as building blocks were constructed and their several properties were investigated [34-36]. As an easy and low cost

technique, it is emerging as a promising approach for the preparation of nanostructures on large areas as the use of different solvents and size nanoparticles results with a variety of morphologies. In order to locate the nanoparticles into the preferred orientations and construct the desired 2D or 3D structures, it is necessary to understand and control the forces governing behavior of nanoparticles in solutions. In this regard, a recent study showed that the arrangement of CdTe nanoparticles into nanowires with the use of dimethyloxide, which enhanced the electrostatic interactions in the solution, is indeed possible [37]. In a following report, Zhang and Wang demonstrated the utility of anisotropic electrostatic interactions of negatively charged gold nanoparticles for their arrangement into chains by modifying the ionic strength and polarity of the aqueous solution. Electrostatic interactions are defined as ubiquitous interactions present in aqueous solutions and they are long-ranged in nature. The attractive and repulsive electrostatic forces are considered as isotropic and the major driving force for the assembly of charged nanoparticles while dipole interactions are short-range in nature. The balance between these forces existing in solution can be utilized for their assembly. Fig. (**1**) shows the assembly of charged particles as the potential change of long-range electrostatic and short-range van der Waals attractions. The authors used the Derjaguin–Landau–Verwey–Overbeek (DLVO) theory to estimate the stability of the charged particles in colloidal suspension system [38]. V_{elec} is the sum of the electrostatic repulsion potential, V_{vdW} is the van der Waals attraction potential, V_{dipole} is the dipolar interaction potential, and $V_{charge–dipole}$ is the charge–dipole interaction potential.

The 14 nm thioglycolic acid (TGA) covered gold nanoparticles (TGA-GNPs) were used to test their model. Since the addition of a small amount of NaCl into the colloidal suspension diminished the V_{elec}, the formation of nanoparticle chains was observed with the addition of NaCl. The equations they derived indicated that a decrease in the dielectric constant of the colloidal suspension containing TGA-GNPs resulted in a decrease in V_{elec} and increase in V_{dipole}. First, they demonstrated the formation of the TGA-GNP chains with the addition of small amount of NaCl, and then, demonstrated the same formation with the addition of polar organic solvents such as acetonitrile, ethanol and acetone, which have smaller dielectric constants. The chain formation by GNPs can be easily observed with UV/Vis spectroscopy because the peak due to longitudinal resonance appears in addition to the peak due to transverse resonance. Fig. (**2**) shows that the UV/V spectrum increased TGA-GNP chain length as the acetonitrile ratio increased as well as the resolution of the TEM images of the TGA-GNP chains.

The spontaneous formation of ordered patterns of nanoparticles using Marangoni flow of ethanol/water mixture and "tears of wine" phenomenon was reported by Cai *et al.* In that study, the ordered hexagonal ring and dotted stripe like pattern of polystyrene nanoparticles on SiO_x substrate were successfully prepared. Fig. (**3**) illustrates the self-assembly of nanoparticles. A nanoparticle/methanol suspension is placed on the substrate as a thin film, and this film is exposed to wet N_2 flow. As volatile ethanol evaporates at air/ethanol/substrate contact line, this region experiences a temperature decrease. This temperature drop causes the moisture in the N_2 gas flow to condense at the contact line. Since the ethanol suspension contains high concentration of nanoparticles, the nanoparticles diffuse into the water droplet in contact with ethanol. After the water is dried, the ordered nanoparticles on the surface are formed [39].

A method to generate lines at centimeter scale

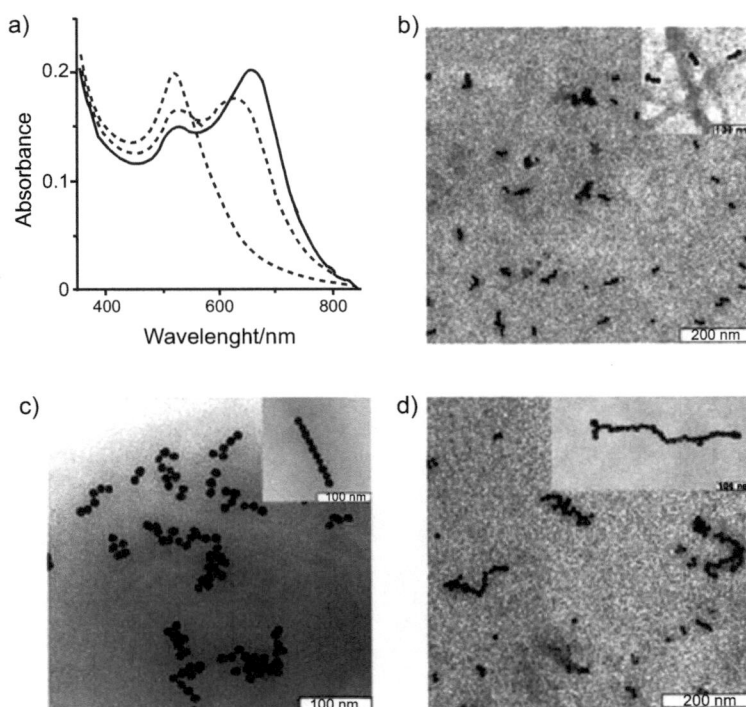

Fig. (2). (**a**) UV/Vis spectra of TGA-GNPs in water–acetonitrile mixture with increasing acetonitrile ratio(1:1, 1:2 and 1:3) . TEM images (**b-d**) of the TGA-GNP chains at the same water–acetonitrile volume ratios (**b**) 1:1, c) 1:2, and d) 1:3). Reprinted with permission from Ref. 38. Copyright @ Wiley-VCH Verlag GmbH & Co. KGaA.

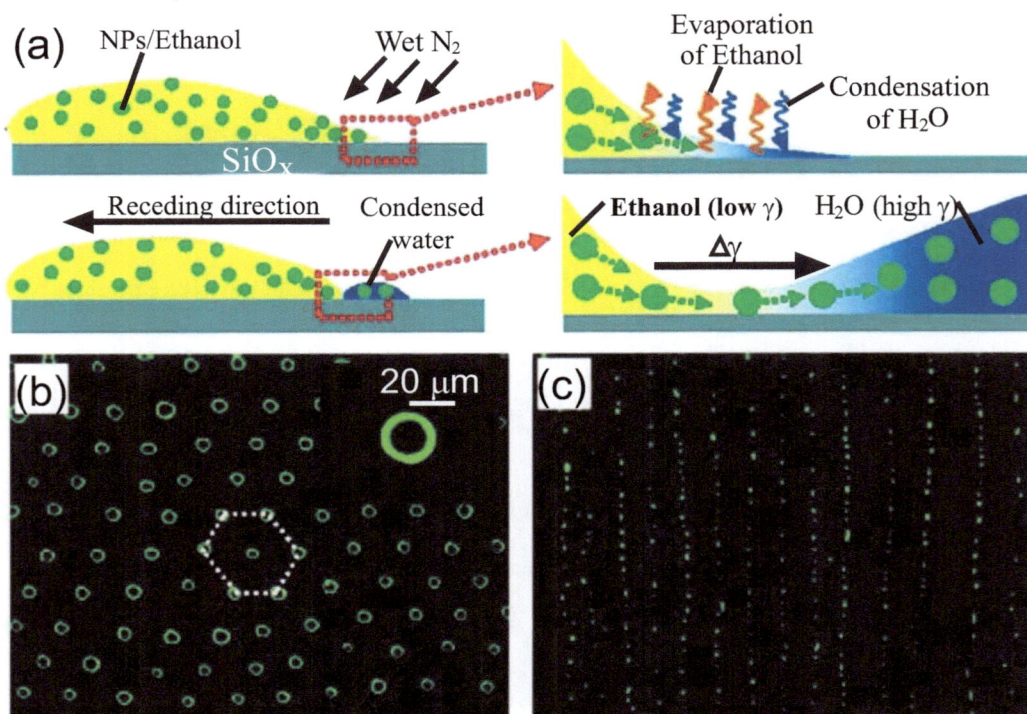

Fig. (3). Illustration of self-assembly of nanoparticles with Marangoni flow method (**a**), fluorescent images ordered hexagonal ring (**b**) and dotted striplike (**c**) patterns on SiO$_x$ substrate. Reprinted with permission from Ref. 39. Copyright @American Chemical Society.

composed of a few nanometers to micrometer sizes of particles was reported [40]. The assembly of nanoparticles was achieved first dispersing the colloidal nanoparticles in an organic solvent such as chloroform and toluene. In the second step, this suspension was dispersed onto the water surface. Upon evaporation of organic solvent, a Langmuir monolayer of particles remains on the water surface. A substrate having medium contact angle is submerged into the water surface possessing the particle monolayer, and is pulled slowly out of the water. The footprints of the semiperiodic "stick and slip" of the contact line assembles the nanoparticles into a line on surfaces.

The unique optical properties of noble metal particles such as gold and silver nanoparticles lead the great interest for their use in sensing [41,42], diagnosis [43], and nanomedicine [44]. The formation of surface plasmons (SPs) because of the interaction of the noble nanoparticles with light offers new opportunities for preparation of novel devices from diagnostics to electronics [43, 45].

The generation and use of this phenomenon is called plasmonics, which is considered as a branch of nanophotonics. The SPs play a significant role in surface-enhanced Raman scattering (SERS) and it is thought that more than 80% of the enhancement mechanism is due to the surface plasmons excited on the noble metal surface. Since the localization and the sensitivity of the SPs strictly depend on the size, shape and position of the nanoparticles relative to each other, it is critically important to organize

nanoparticles into well-defined structures in a reproducible manner. Therefore, it is our ongoing effort to construct such structures for reproducible and sensitive SERS measurement. We have recently developed an approach to assemble the silver nanoparticles into about 1-micrometer size aggregates [46]. In order to prepare such aggregates, a hydrophobic surface prepared by depositing dicholoromethylsilane on an ordinary glass slide and a concentrated colloidal silver nanoparticle suspension was potted. During the evaporation of water from the colloidal suspension, the micrometer size aggregates were formed. The reasonable high SERS enhancement and highly reproducible SERS spectra were obtained from the individual aggregates. Fig. (**4**) shows the SEM images of such aggregates. Besides, the addition of NaCl or SDS into the colloidal suspension influenced the silver nanoparticle spacing as evidenced from the SERS spectra. The assembly of the silver nanoparticles into thin films with the nanometer size cavities from a drying droplet was also demonstrated [47]. The nano-cavities in the thin film were generated by forming a double layer of CTAB molecules before the assembly process. The prepared thin films showed a very high SERS enhancement due to the formation of 2-4 nm gaps between silver nanoparticles, which is necessary for optimal SERS enhancement.

In a separate study, we successfully assembled the bacterial cells and silver nanoparticles on ordinary glass surfaces using "convective assembly" method [48]. In a SERS experiment, molecules and

Fig. (4). SEM image of silver nanoparticle aggregates assembled on a hydrophobic surface from its colloidal droplets.

molecular structures must be as close as possible to the noble metal nanoparticles and surfaces to experience the SPs. When the size of the molecular organization such as bacterial cells is much larger than nanoparticle size, the bacterial cells cannot come to contact with nanoparticles from all sides and SERS spectra obtained from such samples show considerable variations, which introduces uncertainties for bacterial identification. Fig. (**5**) shows the SEM image of the assembled S. cohnii and silver nanoparticles.

4. TEMPLATE ASSISTED SELF-ASSEMBLY

The template assisted or directed assembly of

nanoparticles is a powerful and well-established technique now and the excellent reviews about the subject can be found in literature for the detailed discussion of the topic [49, 50]. In this approach, the templates are generated with a physical or chemical means such as a lithographic technique; e-beam lithography, light lithography, a co-polymer and chemical functionalization. The nanoparticles can be physically directed into the holes or groves with coating techniques such as "convective assembly", dip- and spin coating [51, 52] or the nanoparticles are covalently or through weak interactions such as hydrogen bonding or hydrophobic interactions attached to the functionalized areas.

There is a considerable effort to find alternative approaches for the high-cost, slow, short-range

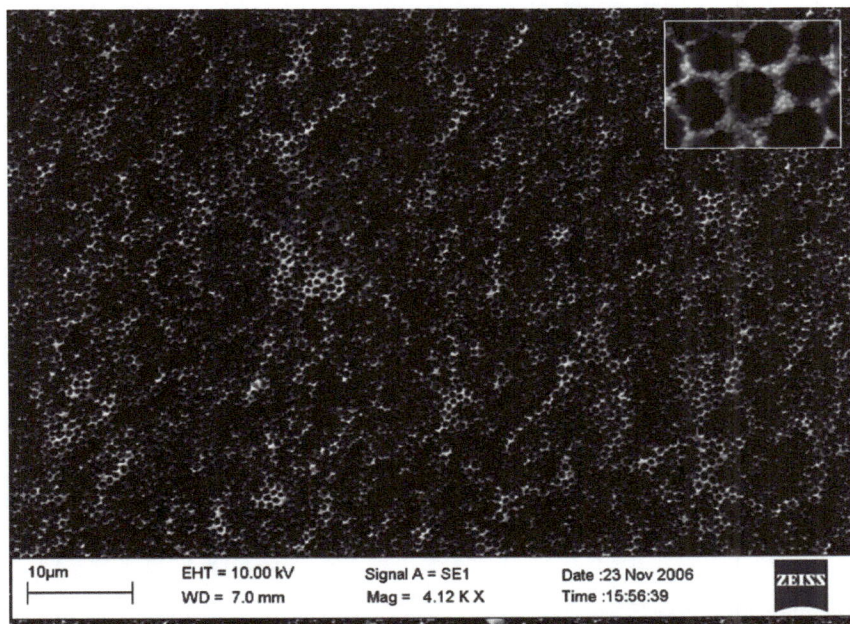

Fig. (5). SEM image of self-assembled bacteria and silver nanoparticles with "convective-assembly" method. Reprinted with permission from Ref. 47. Copyright @American Chemical Society.

lithographic techniques. In this regard, the assembly of micropheres atop nanospheres using colloidal lithography and two-step self-assembly to generate hierarchical nanoparticle pattern arrays was described as low-cost and simple approach [53]. A nanoparticle film composed of 15-50 nm nanoparticles is prepared as a thin film on a substrate. In the second step, the large microparticles are assembled onto the nanoparticle thin films to form well-ordered patterns. In the third step, the voids of the upper microparticle layer were etched with reactive etching method. In the final step, the remaining of the upper layer was removed and the hierarchical nanoparticle pattern was obtained. The authors reported that the hexagonal shaped pattern was easily obtained. The shape and the diameter of the nanoparticle film disks depend on the etching time while the periodicity of the pattern remained the same.

A lithography-free approach for directed colloidal self-assembly based on the wrinkles generated on PDMS was reported [54]. In order to generate wrinkles, poly (sodium 4-styrene sulfonate) and poly (allylamine hydrochloride) (PAH-PSS) thin films were prepared by spraying onto PDMS and a stretch-retraction process was applied to form wrinkled surfaces. Finally, the colloidal particles were deposited into the groves with dip-coating process. The assembled structure using this method was quite well ordered and defect-free.

The use of block copolymers is promising to overcome the difficulties of the current patterning approaches since highly resolved defect free patterns on extended areas with a low cost can be prepared. With the application of block copolymers, 10-30 nm size features can currently be achieved depending on the molecular weight of the polymers used [55-60]. Although the block copolymers can be easily prepared under the defined conditions, there are still certain problems that must be addressed. The challenges are the diversity of the patterns that should be produced and the range of area where the pattern must be prepared. To overcome these problems, Craig and co-workers reported a fabrication approach for well-ordered square arrays of sub–20-nm features without using chemical pattern on the surfaces [61]. In order to prepare complex structures, the same group has investigated the utility of A-B-C triblock polymers other than the A-B diblock polymers. It was possible to enhance the physical properties and broaden the processing window; the use of different blends of polymers

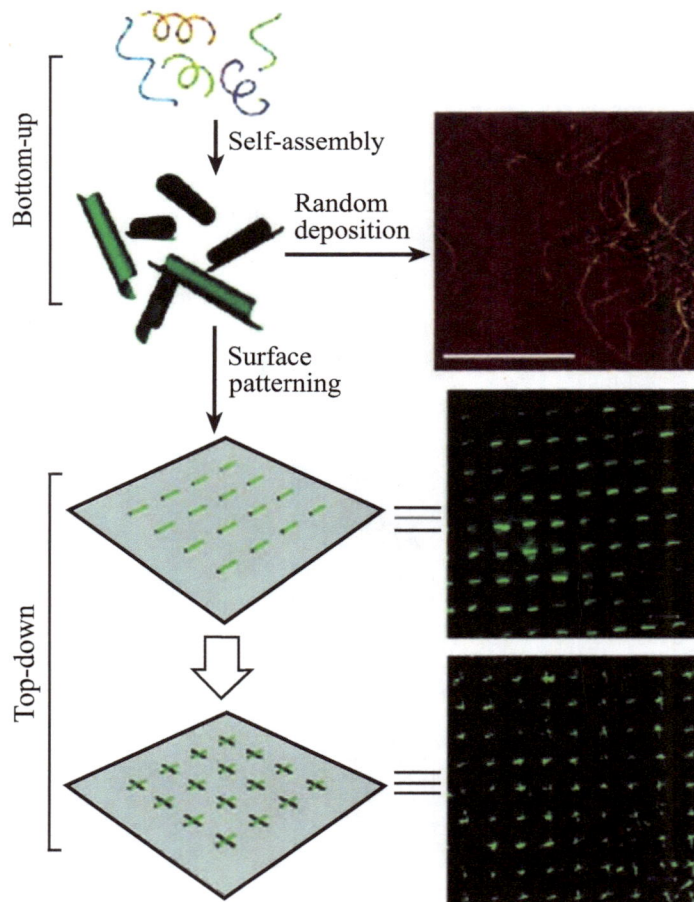

Fig. (6). Illustration of the preparation of DNA nanotube arrays on surfaces by combining "bottom-up" and "top-down" approaches. The images on the right hand side are AFM (top) and fluorescence microscopy images (bottom two). Reprinted with permission from Ref. 97. Copyright @ Wiley-VCH Verlag GmbH & Co. KGaA.

such as A-B/B-C and A-B/C-D alloys could be advantageous [62-64]. However, the difficulties for the generation of uniform thin films have been reported due to the macro phase separation [65-68]. This latest report claims the overcoming such a problem by utilizing the H bonding in addition to the nonspecific dispersive interactions. The control of the composition of H bonding units, block copolymers, molecular weights, and the relative amounts of polymers composing the block copolymer allows the preparation of the patterns with tunable dimensions [69]. Using this approach, they were able to prepare the square arrays of 20-nm cylindrical pores.

There are also other reports addressing the problems associated with the preparation of the template patterns to improve the copolymer directed self-assembly [70]. A topographical graphoepitaxy technique that allowed the assembly of block copolymer thin films on larger areas with high ordered profiles was also described [71].

5.   PROGRAMMED SELF-

ASSEMBLY

The use of DNA as a guide to assemble nanoparticles to generate periodic assembly on surfaces using the well-known hybridization of DNA base pairs is an elegant approach and metal nanoparticles [72-75] and proteins [76-80] can be assembled into complex structures using the DNA as a platform. There are a number of reports about its use for patterning nanoparticles [81-88]. The construction of pre-programmed DNA "smart tiles" has recently been demonstrated [89-95]. In a recent report, CdSe/ZnS quantum dots (QDs) were assembled into hierarchical order by utilizing the DNA as a template. A set of double-crossover (DX) molecules called ABCD tile system was used [96]. The biotin molecules are attached to one of the tiles (Tile A) through a short DNA stem. When the tile system is formed, the four-tiled system gives 2D arrays forming parallel lines of the biotin molecules. The distance between two neighboring biotin lines was estimated to be 64 nm and the distance between biotin molecules on the same line is estimated as 4-5 nm from the AFM measurements. With the addition of streptavidin-attached CdSe/ZnS-QDs onto the

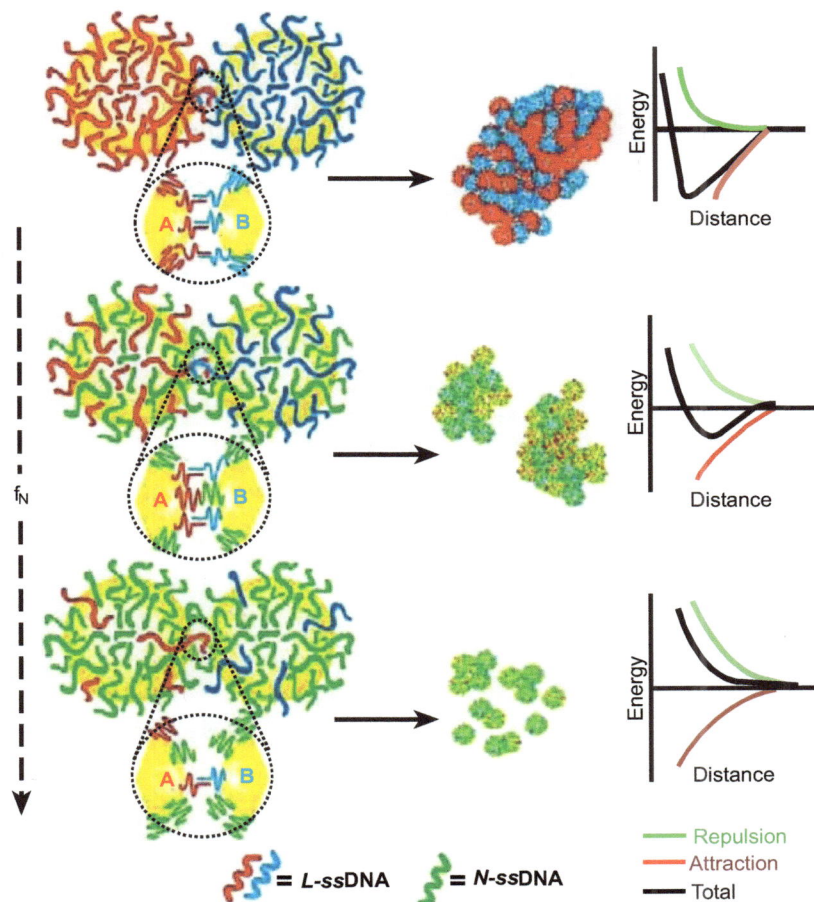

Fig. (7). Illustration of nanoparticles modified with oligonucleotides. The energy diagrams are seen on the side illustrating the balance between attractive and repulsive forces. Reprinted with permission from Ref. 98. Copyright @ Wiley-VCH Verlag GmbH & Co. KGaA.

array, QDs were organized on the tile system thought the biotin-straptavidin specific binding. This study indicated that QD-array could be systematically prepared on surfaces [97].

Lin *et al.* adapted an approach that combines the bottom-up and top-down approaches to assemble DNA nanotubes on surfaces [98]. First, DNA nanotubes were assembled from a group of rationally designed oligonucleotides. Then, the DNA nanotubes were transferred onto glass surfaces using micrometer size patterned polydimethylsiloxane (PDMS) stamp. Fig. (**6**) shows the construction process of DNA nanotubes on surfaces. The

oligonucleotides used to prepare the DNA nanotubes can be easily modified with several functional groups such as thiol, amino, carboxylic group, etc., which allows the covalent attachment of several nanoparticles or biomolecules such as antibodies and biotin, to construct variety of complex patterned structures.

In addition to the programmed assembly at surfaces, the DNA hybridization based assembly can be performed in solution to obtain aggregates with well-defined morphology. The molecular recognition and steric repulsion forces were used to control the assembly morphology [99]. The 1.9 μm

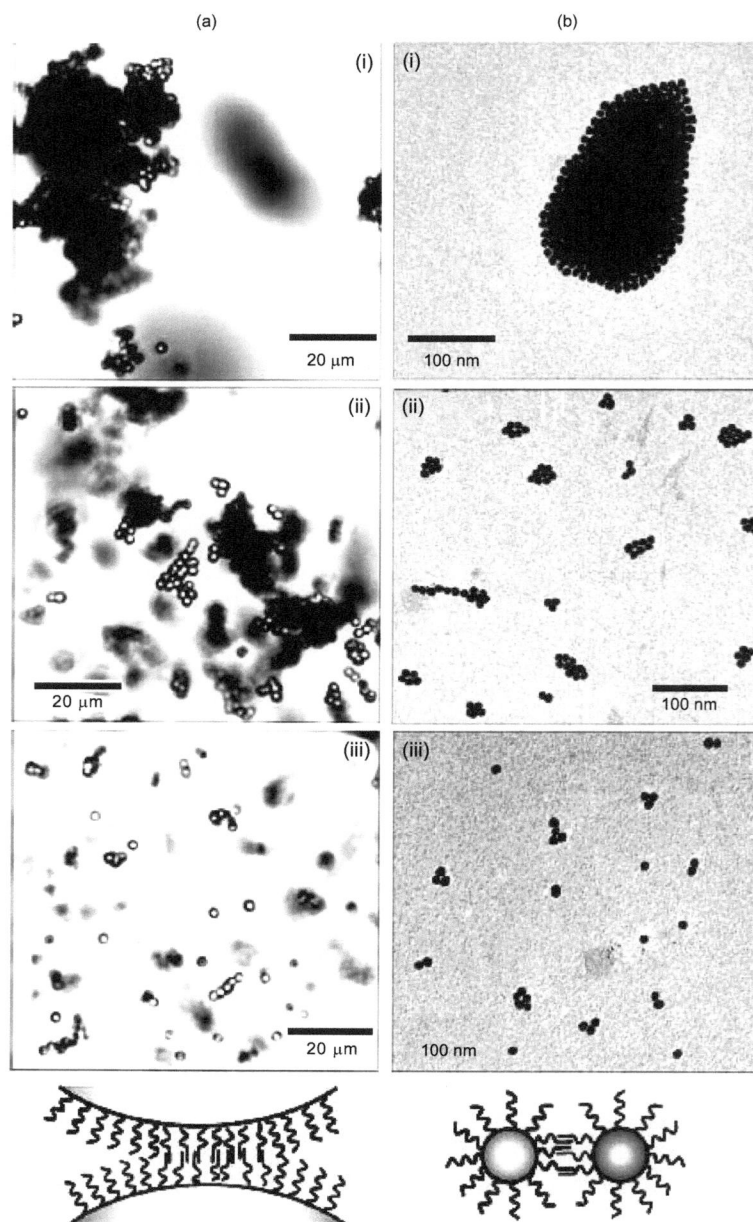

Fig. (8). Optical microscopy images of the system prepared with PS particles (a) and TEM images of the system prepared with GNPs (b). For PS-system with f_N of i) 0.0, ii) 0.7, and iii) 0.9 [[A]=[B]=0.05% w/v, PBS buffer, pH 7.4) and for GNP-system with f_N of i) 0.75, ii) 0.85, and iii) 0.95 ([A]=[B]=7.5 nm, 10mm phosphate buffer, 0.3m NaCl, pH 7.1). Reprinted with permission from Ref. 98. Copyright @ Wiley-VCH Verlag GmbH & Co. KGaA.

polystyrene particles and 9.6 nm gold nanoparticles were assembled with the balance between hybridization of the complementary DNA and repulsion between non-complementary oligonucleotides. The interaction between nanoparticles was regulated with complementary and non-complementary oligonucleotides which are seen in Fig. (**7**). The 1.9-mm-diamater polystyrene particles and 9.6-nm-diameter gold particle systems composed of an equimolar mixture of two types of particles (A and B) were used. The fraction of oligonucleotide attached to surface of the nanoparticles was defined as $f_N=[N]/([N]+[L]$, where L is the composition of the complementary linker ssDNA and N is noncomplex-mentary ssDNA. By changing the f_N, it was possible to control the aggregation morphology. Fig. (**8**) shows the optical microscopy and TEM images of DNA covered PS and GNPs with increasing f_N.

In two separate recent studies, DNA-guided crystallization of gold nanoparticles was investigated [100, 101]. In previous studies, the most of DNA-mediated assembly experiments resulted in amorphous polymeric structures [102-107]. However, these two latest studies demonstrated that when the DNA coated nanoparticles were crystallized using different DNA sequences; the formed crystal structure showed different crystalline states of the same inorganic gold nanoparticles. In the latter study, it was demonstrated that the formation of the nanoparticle crystals from DNA coated gold nanoparticles was reversible during heating and cooling cycles.

In a similar way to DNA-based assembly, rationally designed peptides can be used for the programmed assembly. A recent report clearly demonstrates the feasibility of this approach by combining the nanoparticle nucleation and growth control property of peptides and peptide self-assembly [108]. The assembly of nanoparticles was achieved with the use of peptide conjugates. A peptide sequence, AYSSGAPPMPPF, confirmed for its gold surface binding property, was used. Since this sequence is composed of hydrophilic amino acids, attachment of a hydrophobic tail that is an aliphatic carbon chain with twelve carbons to N-terminus of the peptide sequence lead the formation of ribbon-like structures appearing to twist in a left-handed direction when chloroauric acid is reduced in the presence of HEPES buffer and the peptide sequence. Fig. (**9**) shows the prepared peptide assisted ribbon-like gold assembly. It should be noted that the reduction of chloroauric acid in the presence of HEPES buffer and the peptide without the hydrophobic tail resulted in monodisperse gold nanoparticle formation.

6.   ASSEMBLY AT LIQUID-LIQUID INTERFACE

The formation of a thin film of micrometer size particles at the interface of two immiscible liquids

was observed decades ago [109, 110]. The behavior of several particles such as iron oxide, silicon dioxide and bariumsulfate in water-paraffin emulsions was studied [111,112]. The recent motivation for the investigation of the behavior of nanometer size particles at liquid interfaces is due to the possibility of preparation of defect-free assembly of nanoparticles on larger areas [113]. Pieranski investigated the basis of the assembly of spherical nanoparticles and he reasoned the behavior of nanoparticles to the decreased total free energy at the oil-water interface [114]. The driving forces for the formation of thin films at interfaces are well understood and also related to the particle size, interaction between the particles, and particle-liquid phase interactions. However, as the particle size gets smaller from micrometer to nanometers sizes, the characterization of nanoparticles at liquid interfaces becomes more difficult and remains to be a challenge. There are excellent reviews regarding the detailed thermodynamic treatment of the behavior of the nanoparticles at liquid interfaces [115,116, 117] and we only focus on the recent application in this chapter.

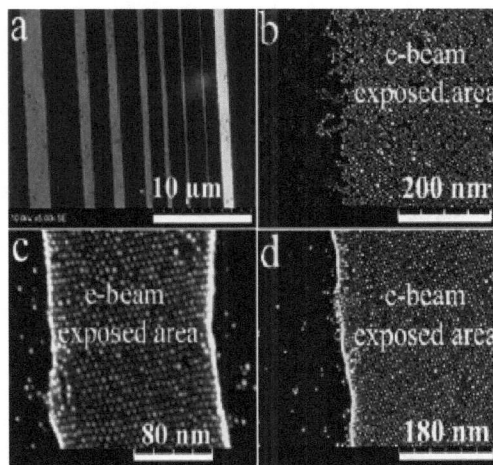

Fig. (9). Electron tomographical data of structure (**a,b**) and schematic depiction of the formation of gold nanoparticle double helices (**c**). Reprinted with permission from Ref. 107. Copyright @American Chemical Society.

The assembly of hydrophobic gold nanoparticles on ordinary substrate surfaces upto 10 cm^2 was successfully achieved at water-toluene interface where toluene contained polymethylmethacrylate (PMMA) [118] in a recent study. As the solvent evaporates, a thin film of gold nanoparticles and polymer composite forms at the interface and it can easily be transferred onto an ordinary substrate. Fig. (**10**) shows thin films prepared on a surface. The gold nanoparticles were further assembled into the patterns by exposing the thin film to electron-beam. The low and high e-beam intensities cause the GNP/PMMA thin film behaves differently. At low e-beam intensities, the GNPs are differentially solubilized in the thin film forming somewhat diffuse interfaces (see Figs. (**11a**) and (**b**)). At high

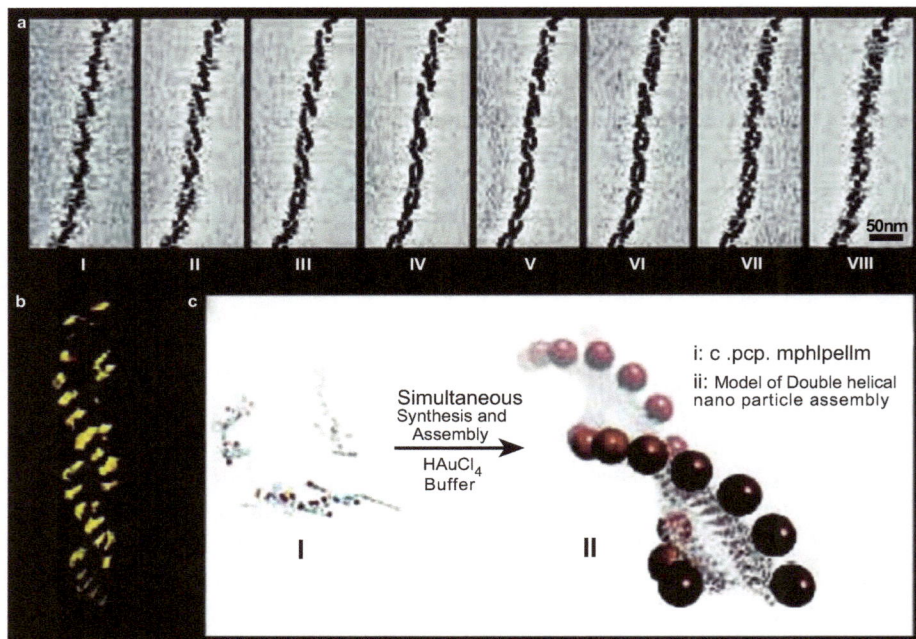

Fig. (10). Light microscopy image of GNP/PMMA film transferred onto glass (**a**) and SEM image of NP/PMMA film transferred to silicon (**b**). Small-angle X-ray scattering pattern of film prepared as in (**b**) (**c**) and TEM of GNP/PMMA film (**d**). Reprinted with permission from Ref. 116. Copyright @American Chemical Society.

e-beam intensities, the PMMA acts as a negative resist and the edge roughness of the formed pattern approaches GNP diameter (see Figs. (**11c**) and (**d**)).

In another recent study, layer-by-layer (LBL) assembly of nanoparticles at water-oil interface was achieved by functionalizing the nanoparticles at the interface [119]. A thin film composed of layers of gold, silver and CdTe nanoparticles was successfully assembled at water-toluene interface. First, DNA bases were functionalized with a spacer molecule that had a thiol group at the free end. This free thiol group binds to nanoparticles and DNA bases are coupled through a well-known DNA base pairing. The authors claim that it is possible to generate large asymmetric multiplayer films composed of different types of nanoparticles utilizing their approach.

7.   SHAPE INDUCED SELF-ASSEMBLY

The shape-induced assembly of nanoparticles could be another approach. Since the different facets of the shaped nanoparticles have different affinities for

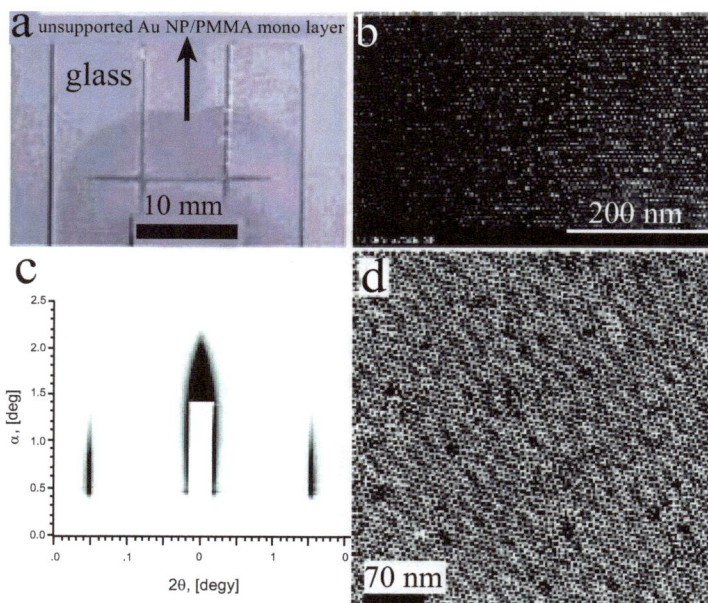

Fig. (11). SEM images of patterns prepared by exposing GNP/PMMA thin films to low (a, b) and high (c, d)e-beam intensities. Reprinted with permission from Ref. 116. Copyright @American Chemical Society.

ligands, this anisotropy can be used to utilize the assembly of nanoparticles because the status of the ligands on the facets of the nanoparticles depends on the crystallographic nature of the facet [120-128]. There are a number of reports utilizing the shape-induced assembly of nanoparticles [129-132]. The early findings indicate that shape induced assembly could be an opportunity to generate structures that are combination of the nanoparticles with different shapes by adding a different nanoparticle to the different facet by utilizing the anisotropy difference among the facets [133-135]. In a recent report, the spontaneous self-assembly of octahedral-shaped silver nanoparticles into three-dimensional plasmonic crystals by utilizing shape-induced assembly was demonstrated. The self-assembly was achieved by the steric repulsive forces due to a polymeric structure used to cover the particle surface. It was observed that the crystallization started in a manner of layer-by-layer formation before the evaporation of all solvents from the suspension.

6. CONCLUSIONS

Here the recent advances in self-assembly of nanostructures are summarized following brief fundamentals. The "bottom-up" is continuously evolving and becoming a versatile tool for the assembly of nanoparticles and nanostructures. As it is clear, the weak forces play dominant role in the self-assembly process and the approaches to control these forces to direct the nanostructures into the desired orientations need to be fully understood. The stage of the assembly techniques and methods developed so far is nowhere close to the desired level yet. Considering the nature's powers and tools to assembly the molecules and molecular structures into higher organizations, human being has to learn whole a lot more than what information is available today. Learning more about nature's machinery will bring more power on our hands to manipulate nanostructures into more sophisticated devices.

7. ACKNOWLEDGEMENTS

The financial support from Yeditepe University Research Fund and TUBITAK is greatly acknowledged for the part of the work presented here.

8. REFERENCES

[1]   Halperin, W. P. Rev. Mod. Phys., 1986, 58, 533.
[2]   Urban, J. J.; Talapin, D. V.; Shevchenko, E. V.; Murray, C. B. J. Am. Chem. Soc., 2006, 128, 3248.
[3]   Giersig, M.; Hilgendorff, M. Eur. J. Inorg. Chem., 2005, 3571.
[4]   Pileni, M. P. J. Phys. Chem. B, 2001, 105, 3358.
[5]   Pileni, M. P.; Lalatonne, Y.; Ingert, D.; Lisiecki, I.; Courty, A. Faraday Discuss, 2004, 125, 251.
[6]   Weller, H. Angew. Chem., 1993, 105, 43; Angew. Chem. Int. Ed., 1993, 32, 41.
[7]   Nirmal, M.; Brus, L. Acc. Chem. Res., 1999, 32, 407.
[8]   Burda, C.; Chen, X.; Narayanan, R.; El-Sayed, M. A. Chem. Rev., 2005, 105, 1025.
[9]   El-Sayed, M. A. Acc. Chem. Res., 2004, 37, 326.
[10]   Eustis, S.; El-Sayed, M. A. Chem. Soc. Rev., 2006, 35, 209.
[11]   Shipway, A. N.; Katz, E.; Willner, I. Phys. Chem. Phys. Chem., 2000, 1, 18.
[12]   Rosi, N. L.; Mirkin, C. A. Chem. Rev., 2005, 105, 1547.
[13]   Yu, B.; Meyyappan, M. J. Solid State Electrochem., 2006, 50, 536.
[14]   Wieckowski, A.; Savinova, E. R.; Vayenas C. G. Catalysis and electrocatalysis at nanoparticle surfaces; Marcel Dekker: New York, 2003.
[15]   Wang, X.; Zhuang, J.; Peng, Q.; Li Y. Nature, 2005, 437, 121.
[16]   Murray, C. B.; Norris, D. J.; Bawendi, M. G. J. Am. Chem. Soc., 1993, 115, 8706.
[17]   Yin, Y.; Alivisatos, A. P., Nature, 2005, 437, 664.
[18]   Klimov, V. I. Semiconductor and metal nanocrystals; Marcel Dekker: New York, 2003.
[19]   Kumar, S.; Nann, T. Small, 2006, 2, 316.
[20]   Bønnemann, H.; Richards, R. Eur. J. Inorg. Chem., 2001, 2455.
[21]   Murray, C. B.; Sun, S.; Doyle, H.; Betley, T. MRS Bull., 2001, 985.
[22]   Pileni, M. P. J. Phys. Chem. B, 2001, 105, 3358.
[23]   Xia, D.; Li, D.; Luo, Y.; Brueck, S. R. J.; Adv. Mater., 2006, 18, 930.
[24]   Wu, X. C.; Chi, L. F.; Fuchs, H. Eur. J. Inorg. Chem., 2005, 3729.
[25]   Hao, E.; Lian, T. Q. Langmuir, 2000, 16, 7879.
[26]   Lu, C. H.; Wu, N. Z.; Jiao, X. M.; Luo, C. Q.; Cao, W. X. Chem. Commun. 2003, 1056.
[27]   Binder, W. H. Angew. Chem., 2005, 117, 5300.
[28]   Zhang, H.; Wang, C. L.; Li, M. J.; Ji, X. L.; Zhang, J. H.; Yang, B., Chem. Mater., 2005, 17, 4783.
[29]   Dubertret, B.; Skourides, P.; Norris, D. J.; Noireaux, V.; Brivanlou, A. H.; Libchaber, A. Science, 2002, 298, 1759.
[30]   Mulder, W. J. M.; Koole, R.; Brandwijk, R. J.; Storm, G.; Chin, P. T. K.; Strijkers, G. J.; de Mello Donega, C. , Nicolay, K.; Griffioen, A. W. Nano Lett., 2006, 6, 1.
[31]   Hao, E.; Lian, T. Q. Langmuir, 2000, 16, 7879.
[32]   Denkov, N. D.; Velev, O. D.; Kralchevsky, P. A.; Ivanov, I. B.; Yoshimura, H.; Nagayama, K. Nature, 1993, 361, 26.
[33]   Kralchevsky, P.; Nagayama, K. Langmuir, 1994, 10, 23.
[34]   Hermanson, K. D.; Lumsdon, S. O.; Williams, J. P.; Kaler, E. W.; Velev, O. D. Science, 2001, 294, 1082.
[35]   Yuan, Z.; Petsev, D. N.; Prevo, B. G.; Velev, O. D.; Atanassov, P. Langmuir, 2007, 23, 5498.
[36]   Prevo, B. G.; Velev, O. D. Langmuir, 2004, 20, 2099.
[37]   Lilly, D.; Lee, J.; Sun K.; Tang, Z.; Kim, K.S.; Kotov, N. J. Phys. Chem. C, 2008, 112, 370.
[38]   Israelachvili, J. Intermolecular & Surface Forces; Academic Press: London, 1997.
[39]   Cai, Y.; and Newby, B. Z. J. Am. Chem. Soc., 2008, 130, 6076.
[40]   Huang, J.; Tao, A. R.; Connor, S.; He, R.; Yang, P. Nano Lett., 2006, 6, 524
[41]   Maier, S. A.; Brongersma, M. L.; Kik, P. G.; Atwater, H. A. Phys. Rev. B, 2002, 65, 193408.
[42]   Maier, S. A.; Kik, P. G.; Atwater, H. A. Appl. Phys. Lett., 2002, 81, 1714.
[43]   Huang, X.; El-Sayed, I. H.; Qian, W.; El-Sayed, M. A. J. Am. Chem. Soc.,2006, 128, 2115.
[44]   Huang, X.; Jain, P. K.; El-Sayed, I. H.; El-Sayed, M. A. Photochem. Photobiol. 2006, 82, 412.
[45]   Ozbay, E. Science, 2006, 311, 189.
[46]   Culha, M.; Kahraman, M.; Tokman, N.; Turkoglu, G. J. Physc. Chem. C, 2008, 112, 10323.
[47]   Kahraman, M.; Tokman, N; Culha, M. ChemPhysChem, 2008, 9, 902 .
[48]   Kahraman, M.; Yazici, M. M.; Sahin, F.; Culha, M. Langmuir, 2008, 5, 894.
[49]   Wang, D.; Mohwald, H. J. Mater. Chem., 2004, 14, 459.
[50]   Gates, B.; Xu, Q.; Stewart, M.; Ryan, D.; Willson, C. G.; Whitesides, G. M. Chem. Rev., 2005, 105, 1171.
[51]   Xia, D.; Brueck, S. R. J. Nano Lett., 2004, 4, 1295.

[52]   Juillerat, F.; Solak, H. H.; Bowen, P.; Hofmann, H. Nanotechnology, 2005, 16, 1311.

[53]   Xia, D.; Ku, Z.; Li, D.; Brueck, S. R. J. Chem. Mater. 2008, 20, 1847.

[54]   Lu, C.; Mohwald, H.; Ferya, A. Soft Matter, 2007, 3, 1530.

[55]   Park, M.; Harrison, C.; Chaikin, P. M.; Register, R. A.; Adamson, D. H. Science, 1997, 276, 1401.

[56]   Kim, S. O. Nature, 2003, 424, 411.

[57]   Ryu, D. Y.; Shin, K.; Drockenmuller, E.; Hawker, T. P.; Russell, T. P. Science, 2006, 308, 236.

[58]   Mansky, P.; Lui, Y.; Huang, E.; Russell, T. P.; Hawker, C. J. Science, 1997, 275, 1458.

[59]   Segalman, R. A.; Yokoyama, H.; Kramer, E. J. Adv. Mater., 2001, 13, 1152.

[60]   Cheng, Y. J.; Ross, C. A.; Thomas, E. L.; Smith, H. I.; Vancso, G. J. Appl. Phys. Lett. 2002, 81, 3657.

[61]   Park, S. M.; Craig, G. S. W.; La, Y. H.; Solak, H. H.; Nealey, P. F. Macromolecules, 2007, 40, 5084.

[62]   Abetz, V.; Goldacker, T. Macromol. Rapid Commun, 2000, 21, 16.

[63]   Vaidya, N. Y.; Han, C. D. Macromolecules, 2000, 33, 3009.

[64]   Mao, H.; Arrechea, P. L.; Bailey, T. S.; Johnson, B. J. S.; Hillmyer, M. A. Faraday Discuss., 2005, 128, 149.

[65]   Asari, T.; Matsua, S.; Takano, A.; Matsushita, Y.; Macromolecules, 2005, 38, 8811.

[66]   Olmsted, P. D.; Hamley, I. W. Europhys. Lett., 1999, 45, 83

[67]   Kimishima, K.; H. Jinnai, H.; Hashimoto, T. Macromolecules, 1999, 32, 2585.

[68]   Jeon, H. G.; Hudson, S. D.; Ishida, H.; Smith, S. D. Macromolecules, 1999, 32, 1803.

[69]   Tang, C.; Lennon, E. M. ; Fredrickson, G. H.; Kramer, E. J.; Hawker, C. J. Science, 2008, 322, 5900.

[70]   Ruiz, R.; Kang, H.; Detcheverry, F. A.; Dobisz, E.; Kercher, D. S.; Albrecht, T. R.; de Pablo, J. J.; Nealey, P. F. Science, 2008, 321, 936.

[71]   Bita, I.; Yang, J. K. W.; Jung, Y. S.; Ross, C. A.; Thomas, E. L.; Berggren, K. K. Science, 2008, 321, 939.

[72]   Le, J.D.; Pinto, Y.; Seeman, N. C.; Musier-Forsyth, K.; Taton, T. A.; Kiehl, R.A. Nano Lett., 2004, 4, 234.

[73]   Zhang, Y.; Liu, Y.; Ke, H.; Yan, H. Nano Lett., 2006, 6, 248.

[74]   Zheng, J.; Constantinou, P. E.; Micheel, C.; Alivisatos, A. P.; Kiehl, R. A.; Seeman, N. C.; Nano Lett. 2006, 6, 1502.

[75]   Sharma, J.; Chhabra, R.; Liu, Y.; Ke, Y.; Yan, H. Angew. Chem., 2006, 118, 744.

[76]   Williams, B. A. R.; Lund, K.; Liu, Y.; Yan, H.; Chaput, J. C. Angew. Chem., 2007, 119, 3111.

[77]   He, Y.; Tian, Y.; Ribbe, A. E.; Mao, C.; J. Am. Chem. Soc., 2006, 128, 12664.

[78]   Liu, Y.; Lin, C.; Li, H.; Yan, H. Angew. Chem., 2005, 117, 4407.

[79]   Malo, J.; Mitchell, J. C.; Venien-Bryan, C.; Harris, J. R.; Wille, H.; Sherratt, D.J.; Turberfield, A. J. Angew. Chem. 2005, 117, 3117.

[80]   Park, S. H.; Yin, P.; Liu, Y.; Reif, J.; LaBean, T. H.; Yan, H. Nano Lett., 2005, 5, 729.

[81]   Le, J. D.; Pinto, Y.; Seeman, N. C. ; Musier-Forsyth, K.; Taton, T. A.; Kiehl, R. A. Nano Lett., 2004, 4,2343.

[82]   Zhang, J.; Liu, Y.; Ke, Y.; Yan, H. Nano Lett., 2006, 6, 248.

[83]   Sharma, J.; Chhabra, R.; Liu, Y.; Ke, Y.; Yan, H. Angew. Chem., 2006, 118, 744. Angew.Chem. Int. Ed. 2006, 45,730 – 735.

[84]   Deng, Z. , Tian, Y.; Lee, S. H.; Ribbe, A. E.; Mao, C. Angew.Chem., 2005, 117, 3648.

[85]   Li, H.; Park, S. H.; Reif, J. H.; LaBean, T. H.; Yan, H. J.Am. Chem. Soc. 2004, 126,418.

[86]   Lee, J.; Wernette, D. P.; Yigit, M. V.; Liu, J.; Wang, Z.; Lu, Y. Angew.Chem., 2007, 119, 9164.

[87]   Aldaye, F.; Sleiman, H. F. Angew.Chem., 2006, 118, 2262.

[88]   Aldaye, F.; Sleiman, H. F. J.Am. Chem. Soc., 2007, 129,4130.

[89]   Seeman, N. C. Nature, 2003, 421, 427.

[90]   Winfree, E.; Liu, F.; Wenzler, L. A.; Seeman, N. C. Nature, 1998, 394, 539.

[91]   Mao, C.; Sun, W.; Seeman, N. C. J. Am. Chem. Soc., 1999, 121, 5437.

[92]   LaBean, T. H.; Yan, H.; Kopatsch, J.; Liu, F.; Winfree, E.; Reif, J. H.; Seeman, N. C. J. Am. Chem. Soc., 2000, 122, 1848.

[93]   Sha, R.; Liu, F.; Millar, D. P.; Seeman, N. C. Chem. Biol., 2000, 7, 743.

[94]   Yan, H.; LaBean, T. H.; Feng, L.; Reif, J. H. Proc. Natl. Acad. Sci. U.S.A., 2003, 100, 8103.

[95]   Yan, H.; Park, S. H.; Finkelstein, G.; Reif, J. H.; LaBean, T. H. Science, 2003, 301, 1882.

[96]   Liu, F.; Sha, R.; Seeman, N. C. J.Am.Chem. Soc., 1999, 121, 917.

[97]   Sharma, J.; Ke, Y.; Lin, C.; Chhabra, R.; Wang, Q.; Nangreave, J.; Liu, Y.; Yan, H. Angew. Chem. Int. Ed., 2008, 47, 5157.

[98]   Lin, C.; Ke, Y.; Liu, Y.; Mertig, M.; Gu, J.; Yan, H. Angew. Chem. Int. Ed., 2007, 46, 6089.

[99]   Maye, M. M.; Nykypanchuk, D.; van der Lelie, D.; Gang, O. Small, 2007, 3, 10.

[100]   Nykypanchuk, D.; Maye, M. M.; van der Lelie D.; Gang, O. Nature, 2008, 451, 549.

[101]   Chad Mirkin Paper

[102]   Park, S. J.; Lazarides, A. A.; Storhoff, J. J.; Pesce, L.; Mirkin, C. A. J. Phys. Chem. B, 2004,108, 12375.

[103]   Biancaniello, P. L.; Kim, A. J.; Crocker, J. C. Phys. Rev. Lett., 2005, 94, 058302.

[104]   Park, S. Y.; Lee, J. S.; Georganopoulou, D.; Mirkin, C. A.; Schatz, G. C. J. Phys. Chem. B, 2006,110, 12673–12681.

[105]   Maye, M. M.; Nykypanchuk, D.; van der Lelie, D.; Gang, O. J. Am. Chem. Soc., 2006,128, 14020–14021.

[106]   Maye, M. M.; Nykypanchuk, D.; van der Lelie, D.; Gang, O. Small, 2007, 3, 1678–1682.

[107]   Park, S. J.; Lazarides, A. A.; Mirkin, C. A.; Letsinger, R. L. Angew. Chem. Int. Edn Engl. 2001, 40, 2909–2912.

[108]   Chen C. L.; Zhang P.; Rosi N. L. J. Am. Chem. Soc., 2008, 130, 13555.

[109]   Pickering S. U. J. Chem. Soc. Trans., 1907, 91, 2001.

[110]   Ramsden W. Proc. R. Soc. London, Ser. A, 1903, 72, 156.

[111]   Pieranski P. Phys. Rev. Lett. 1980, 45, 569.

[112]   Binks B. P.; Lumsdon S. O. Langmuir, 2000, 16, 8622.

[113]   Pang J.; Xiong S.; Jaeckel F.; Sun Z.; Dunphy D.; Brinker C. J. J. Am. Chem. Soc.,2008, 130, 3284

[114]   Bresme F.; Oettel M. J. Phys. Condens. Matter, 2007, 19, 413101

[115]   Boker A.; He J.; Emrick T.; Russell T. P. Soft Matter, 2007, 3, 1231.

[116]   Bresme1, F.; Oettel, M. J. Phys.: Condens. Matter., 2007, 19, 413101.

[117]   Kinge, S.; Crego-Calama, M.; Reinhoud, D. N. ChemPhysChem 2008, 9, 20–42.

[118]   Pang J., Xiong S.; Jaeckel F.; Sun Z.; Dunphy D.; Brinker C. J. J. Am. Chem. Soc., 2008, 130, 3284.

[119]   Wang B.; Wang M.; Zhang H.; Sobal N. S.; Tong W.; Gao C.; Wang Y.; Giersig M.; Wang D.; Möhwald H. Phys Chem Chem Phys., 2007, 9, 6313.

[120]   Talapin, D. V.; Shevchenko, E. V.; Murray, C. B.; Titov, Al. V.; Kral, P.; Nano Lett., 2007, 7, 1213–1219.

[121]   Tang, Z. Y.; Ozturk, B.; Wang, Y.; Kotov, N. A. J. Phys. Chem. B, 2004, 108, 6927–6931.

[122]   Volkov, Y.; Mitchell, S.; Gaponik, N.; Rakovich, Y. P.; Donegan, J. F.; Kelleher, D.; Rogach, A. L. ChemPhysChem, 2004, 5, 1600–1602.

[123]   Jackson, A. M.; Myerson, J. W.; Stellacci, F. Nat. Mater., 2004, 3, 330–336.

[124]   Polleux, J.; Pinna, N.; Antonietti, M.; Hess, C.; Wild, U.; Schlegl, R.; Niederberger, M. Chem. Eur. J., 2005, 11, 3541–3551.

[125]   Zhang, Z. P.; Sun, H. P.; Shao, X. Q.; Li, D. F.; Yu, H. D.; Han, M. Y. Adv. Mater., 2005, 17, 42–47.

[126]   Tang, Z., N.; Kotov, A.; Giersig, M. Science, 2002, 297, 237–240.

[127]   Tang, Z. Y.; Wang, Y.; Sun, K.; Kotov N. A. Adv. Mater., 2005, 17, 358– 363.

[128] Cho, K. S.; Talapin, D. V.; Gaschler, W.; Murray, C. B. J. Am. Chem. Soc., 2005, 127, 7140–7147.

[129] Tian, Z. R.; Liu, J.; Xu, H.; Voigt, J. A.; Mckenzie, B.; Matzke, C. M.; Zhang, J.; Liu, H.; Wang, Z.; Ming, N., Nano Lett., 2003, 3, 2,179-182.

[130] Zhang, J.H.; Liu, H.Y.; Zhan, P.; Wang, Z.L.; Ming, N.B. Adv. Func. Mater., 2007, 17, 9,1558-1566.

[131] Tian, Z.R.R.; Liu, J.; Xu, H.F.; Voigt, J.A.; Mckenzie, B. Matzke, C.M. Appl. Phys. Lett., 2007, 91, 13, 133112.

[132] Tao, A.R; Ceperley, D. P.; Sinsermsuksakul, P.; Neureuther, A. R.; Yang, P. 2008, in press, ASAP article.

[133] DeVries, G. A.; Brunnbauer, M.; Hu, Y.; Jackson, A. M.; Long, B.; Neltner, B. T.; Uzun, O.; Wunsch, B. H.; Stellacci, F. Science, 2007, 315, 358–361.

[134] Kudera, S.; Carbone, L.; Casula, M. F.; Cingolani, R.; Falqui, A.; Snoeck, E.; Parak, W. J.; Manna, L. Nano Lett. 2005, 5, 445–449.

[135] Cozzoli, P. D.; Manna, L. Nat. Mater. 2005, 4, 801–802.

CHAPTER 6

RATIONAL SYNTHESIS APPROACHES TO METAL NANOPARTICLES AND POLYMER METAL NANOCOMPOSITES

K. Samba Sivudu, [1] Y. Murali Mohan, [1,2] and K. Mohana Raju[1*]

[1]*Department of Polymer Science & Technology, Sri Krishnadevaraya University, Anantapur -515 003, India;*
[2]*Cancer Biology Research Center, Sanford Research/ USD, Sioux Falls, SD-57105*

Address correspondence to: K. Mohana Raju, Department of Polymer Science & Technology, Sri Krishnadevaraya University, Anantapur -515 003, India; Email: kmrmohan@yahoo.com

Abstract: The scientific and industrial need for novel composite materials and nanoparticles has opened new paths and led to significant advances in the field of nanocomposites. A number of nanosystems can be designed based on their unique physico-chemical structures for direct biomedical applications. This review is focused on the most novel strategies and trends to design metal nanoparticles including thermal decomposition, chemical reduction and green methods. These methods can be applied to fabricate polymer metal nanocomposites into different forms: core –shell, hollow core- shell nanoparticle system, metal nanoparticles in hydrogel matrix and layer by layer assembly systems. The resulting new class of materials found fascinating interest in antimicrobial, drug delivery and in catalysis.

Key words: Metal nanoparticles, polymer metal nanocomposites, chemical reduction, green method, core –shell, drug delivery, catalysis.

1. INTRODUCTION

Nanostructured metal and metal oxide-polymer nanocomposites are the subject of great interest to researchers in different areas of science and technology. Such composites, mixed at the molecular level are much different from the conventional composites with incorporation of a variety of additives in the polymer matrices [1]. In polymer metal nanocomposites, strong chemical bonds (covalent or ionic) or interactions such as van der Waals forces, hydrogen bonding, or electrostatic forces, may exist between the organic and inorganic components. This leads to improved physico-chemical properties which may extend for their potential applications in the fields of optics [2], electrical devices [3], mechanics [4], photoconductors [5] and so on. Therefore, recent research has focused on nanocomposites both from fundamental and application points of view [6, 7]. Significant efforts have been paid on the development of novel methods to prepare new hybrids with desired properties and functions [8].

Increasing interest has been focused on the preparation of composite materials which consists of polymer shells encapsulating different core chemical compositions. These powders or particles covering inorganic materials have been used as beads for gas separation, catalysts, coatings, flocculants, toners, raw materials recovery, drug delivery and for anticorrosion protection. In this the selection of the polymeric matrix is crucial for the optimization of systems. For this purpose, the amphiphilic polymeric systems, ranging from nonionic polymers, polyelectrolytes to amphiphilic diblock copolymers are receiving increased attention, since they offer tremendous options for the development of composites possessing unique catalytic, optical, electronic, and magnetic properties.

2. METAL NANOPARTICLES

It is well known that polymers are being employed as protective layer on metal nanoparticles to stabilize the metal nanoparticles (MNP). Mostly, polymers that are surrounded on the metal nanoparticles not only influence the overall stability and dispersion but also enhance their material processability. Further, these polymers may provide as means to finely-tune and tailor the metal nanoparticles in their size and shape for specific needs and applications. Sometimes the experimental condition also influences the size, morphology, stability, the chemical and physical properties of the metal nanoparticles. The kinetics of interaction of metal ions with stabilizers in the presence of reducing agents, and adsorption processes of stabilizing agent with metal nanoparticles are discussed clearly. Generally, metal nanoparticles can be developed by different techniques such as thermal decomposition, soft chemical, green (*insitu*) and radiation methods.

2.1. Thermal Decomposition

Thermal decomposition of volatile metal salts or metal compounds in a polymer matrix is considered to be the best method for the formation of zerovalent metal/

metal oxide particles which can be homogeneously dispersed in the polymer matrix. The set-up consists of four principle units: the zone of treating the polymer precursor, the zone of evaporating metals, the reactor, and the deposition zone. The fluxes of metal atoms (Pd; Sn; Cu, Pd) are evaporated from bulk samples which condense onto the substrate together with the monomers. The condensate consists of nanoparticles of the metal and the monomer. During the heating process of substrate, the monomers are polymerized to produce poly(p-xylylene) [9a]. The structure thus obtained had porous matrix with dispersed nanoparticles in it. Another way of producing metal polymer composites in thermal decomposition route is combination of cobalt and iron carbonyls in the presence of polymers containing $-NH_2$, or $-OH$ functional groups, or unsaturated groups [9b, 10]. The size of the metal nanoparticles formed in such polymer matrices depends on the reaction conditions (temperature, the type of polymer, metal salt and solvent). In addition to the above conditions, the number of polar groups in the polymer backbone seems to be important in controlling the particle size i.e., more polar groups lowers the size of particles [11]. Instead of the metal carbonyls, other organometallic species can also be used as precursors for the metal species in the composite materials. A combination of di-block copolymer and thermal decomposition are used to prepare metal nanoparticles to attach selectively [12-17].

Metal complexes used in the self-assembled microdomain structure of block copolymers were reduced by thermal treatment, and consequently nanoclusters were spatially oriented in the films. Hashimoto *et al.* [15] described the metal nanoparticles formation in the presence of amphiphilic block copolymers solution. In this process, colloidal silver was introduced into microphase of poly(styrene-*b*-2-vinyl pyridine) (PS-b-PVP) diblock copolymer film by the reduction of silver iodide. No silver nanoparticles were found in poly(styrene) phases. Block copolymers may contain nanodomain structures, such as lamellar, cylindrical, or spherical structures,

which can be used to arrange ordered structures of metal clusters. Fig. (1) shows the transmission electron microscopic (TEM) images of the cross section of symmetric PS-b-PMMA film, having the number-average molecular weight of (Mn) 143500, after exposure to the PdIIAA vapour for 30 min.

The lamellar microdomain structure of PS-b-PMMA, where PS and PMMA phases are alternately aligned, can be visualized without heavy metal staining. This indicates that the PdIIAA vapour condensed in one of the phases in the block copolymer. Lee et al [16] also discussed the distribution of metal particles incorporated into crystalline polymers. Poly(t-butyl methacrylate) , atactic polystyrene, polyamide 6(PA6), poly(ethylene terephthalate) and syndiotactic polystyrene(S-PS) films using the one-step dry process in which, the Pd nanoparticles were incorporated into the polymer films via a dry route by a metal complex process without the aid of any solvent or reducing agent. The Pd nanoparticles were successfully incorporated into the crystalline polymers having melting temperatures higher than the processing temperature (180˚C). Here in this process, the Pd nanoparticles were selectively located in the amorphous regions between the lamellae. Therefore, the current method helps to introduce metal nanoparticles into the polymer films with poor solubility and high melting temperatures. The TEM micrographs of PA6 and PET exposed to $Pd(acac)_2$ vapor for 30 min [17]. Both polymers exhibited uniform distribution of Pd nanoparticles in the films. The average diameter of the Pd nanoparticles is in between 3.4 and 3.7 nm with standard deviation of 0.9 and 1.4 nm in PA6 and PET, respectively, which are smaller than those in PS.

Iron (Fe) particles can also be obtained by thermolysis and photolysis of carbonyl complexes. Polymer composites with several transition metal nanoclusters were obtained by decomposing organometallic precursors in phase-separated block copolymers [18-20]. Wojciechowska *et. al.* reported the preparation of gold nanoparticles by decomposition of an

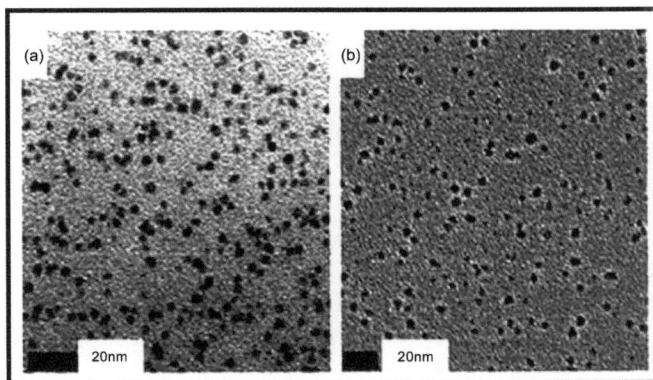

Fig. (1). TEM images of PS-b-PMMA films containing silver nanopticles. Images reprinted with permission from Ref. [15] Copyright © 1999 American Chemical Society.

organometallic precursor dispersed in solid polymer films or as microcrystals [21]. The organometallic complex, 3-oxo[tris(triphenylphosphine)gold](1+)tetra-fluoroborate(1-) [O-(Au(PPh$_3$))$_3$][BF$_4$], was chosen as the Au atom precursor, because it has been shown to be a good source of Au for NP synthesis in solution, due to its more solubility in many organic solvents, and also sufficiently stable to handle in an ambient atmosphere [21].

2.2. Soft Chemical Method

Soft chemical method is used to obtain polymer stabilized metal nanoparticles and polymer-immobilized metal nanocomposites in which the reduction processes are carried out in the form of solutions or suspensions in the presence of a polymer stabilizer, or mononuclear metal complexes bonded to the polymers. Chemical reduction can be carried out by reducing agents such as hydroquinone, phenylenediamine [22], sodium cyanoborohydride and *N,N,N,N*-tetramethyl-*p*-phenylenediamine [23], sodium borohydride [24] hydrazine hydrate [25] (Scheme 2), sodium formaldehyde sulfoxylate [25], ascorbic acid [26], glycerol [27, 28], and DMF [29].

Silver/poly(vinyl alcohol) nanocomposites were prepared through reduction of silver salt by employing two different reducing agents, namely, hydrazine hydrate and sodium formaldehyde sulfoxylate (SFS) . First time, Khanna *et al.* [25] reported the formation of silver nanoparticles by using SFS in aqueous PVA, which is a well-known reducing agent. This mild reducing agent was suitable for obtaining mono disperse particles in the polymer matrix. SFS is highly effective as a reducing agent for the preparation of silver nanoparticles and much easier to handle due to its weak reducing nature as compared to hydrazine hydrate. It also dictates the control of rate of reduction during the formation of silver nanoparticles. The obtained golden-yellow solutions were stable over a long period of time thereby indicating that the nanoparticles have no tendency to agglomerate. Since, PVA acts as a protective agent and restricts the mobility of silver ions during the reaction; it avoids agglomeration.

The ultrafine silver powder was prepared with chemical reduction method by employing ascorbic acid, as a reducing agent. The reaction of AgNO$_3$ with ascorbic acid gives polyhedron monodispersed ultrafine silver nanosystems. The average particle size reduced linearly from 3.1 to 1.0 μm as the reaction temperature increases from 20°C to 60°C. In addition, the existing state of silver ion greatly depends on the pH value in aqueous solution. In fact, the average particle size reduced with increase of pH of reaction medium because the reducing efficacy of ascorbic acid is pH-dependent. It was also noticed that ultrafine silver powder of 1.8 ± 0.4 μm was formed in the absence of any other reducing agent.

Nanosized uniform silver powders and colloidal dispersions of silver were prepared from AgNO$_3$ by a chemical reduction method involving the intermediate preparation of Ag$_2$O colloidal dispersion in the presence of sodium dodecyl sulfate CH$_3$(CH$_2$)$_{11}$OSO$_3$Na as a surfactant. Ascorbic acid was chosen instead of hydrazine hydrate, formaldehyde (HCHO) and glucose (C$_6$H$_{12}$O$_6$) because of its moderate reduction potential. Optimum modes of synthesis were established in the formation of colloidal dispersion of Ag$_2$O particles by ascorbic acid in the presence of sodium dodecyl sulfate. The conditions for the preparation of silver powder with particle sizes in the range 60–120 nm and colloidal dispersions with average size of silver particles 10–50 nm and concentration 0.5–2 wt.% were established [30].

Gao *et al.* [31] developed silver nanodecahedrons first time by liquid-phase reduction of AgNO$_3$ with assistance of poly(vinyl pyrrolidone) (PVP). Ag tetrahedrons dominated in the resulting solution when the ratio between PVP and AgNO$_3$ increases. Finally, the edge size of Ag nanostructures was changed little during the transformation from tetrahedrons to decahedrons and icosahedrons. A typical SEM image of the Ag decahedrons is shown in Fig (**2**).

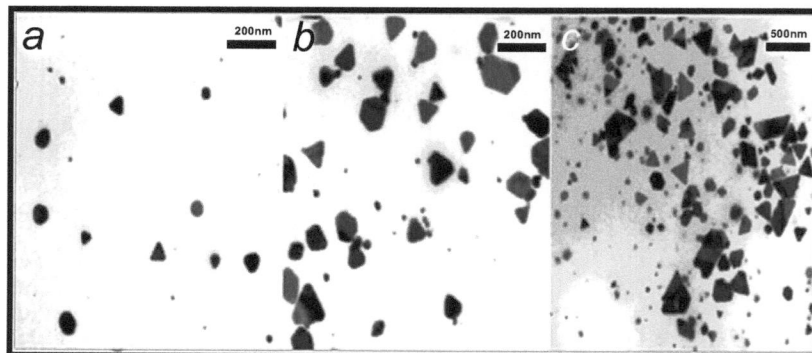

Fig. (3). Silver nanoparticles and nanoprisms formation in presence of PVP in DMF. Images reprinted with permission from Ref. [29] Copyright © 2006 American Chemical Society.

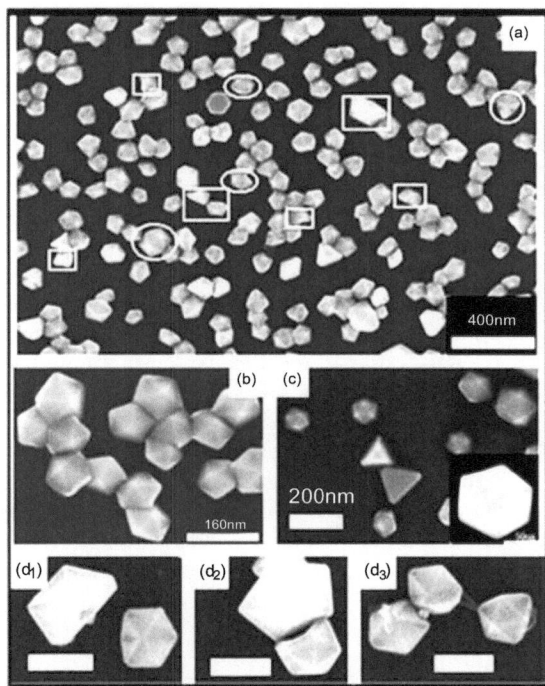

Fig. (2). Nanodecahedrons formation from silver nitrate solution in presence of poly(vinyl pyrrolidone). Images reprinted with permission from Ref. [31] Copyright © 2006 Elsevier.

It was estimated from the SEM image that about 70% of the whole Ag nanoparticles holds decahedronal shape with edge size of 80 nm along with few Ag icosahedrons, Ag nanoprisms and nonoregular nanoparticles. Fig. (**3b**) shows the SEM image of the Ag decahedrons with higher magnification. It seems that the Ag decahedrons were laid on the Si (1 0 0) substrate with various poses. The bright vertexes and five sharp edges of each Ag decahedron are observed distinctly. A series of intermediates of Ag decahedrons and icosahedrons are further illustrated in Fig. (**3d**). The reduced additional rate of DMF solution of PVP K85-95 AgNO3 results in an increase of Ag

icosahedrons Fig. (**3c**). It was known that an icosahedron consists of 20 tetrahedrons sharing one vertex, as shown in the inset of Fig. (**3c**) exhibiting one typical SEM image of Ag icosahedron with higher magnification. Pastoriza *et al.* [29] generated silver nanoprisms by boiling the silver nitrate dissolved in (N,N'-dimethy formamide) (DMF) solvent in the presence of PVP. The size of nanoprisms was controlled with reaction time and temperature. Electron microscopy observation indicates that initially small spheres are formed which then assemble and a melting like process takes place, which leads to crystalline particles with well-defined shapes. The formed nanoprisms become larger with time, and a wider variety of shapes for longer boiling times, are shown in Fig. (**3**) [29].

Morphological changes in silver nanoparticles were observed by the effect of sodium citrate, citric acid and irradiation of light in the presence of PVP by NaBH$_4$ Fig. (**4**).

When citric acid was added to the solution instead of sodium citrate the silver nanoparticles transformed into triangular shape (Figure 5(c)). These triangular silver particles again changed into silver nanorods and triangular shape by irradiation (Figure 5(d)) [32]. Silver dendrites were successfully prepared in DMF solution containing PVP as a stabilizer under microwave irradiation [33].

2.3. Green Methods

Over the past decade, increasing awareness about global warming has led researchers to focus on 'green chemistry'. Utilization of nontoxic chemicals, environmentally benign solvents and renewable materials are some of the key issues that merit important consideration in green synthetic strategies. These efforts are aimed at the total elimination or at

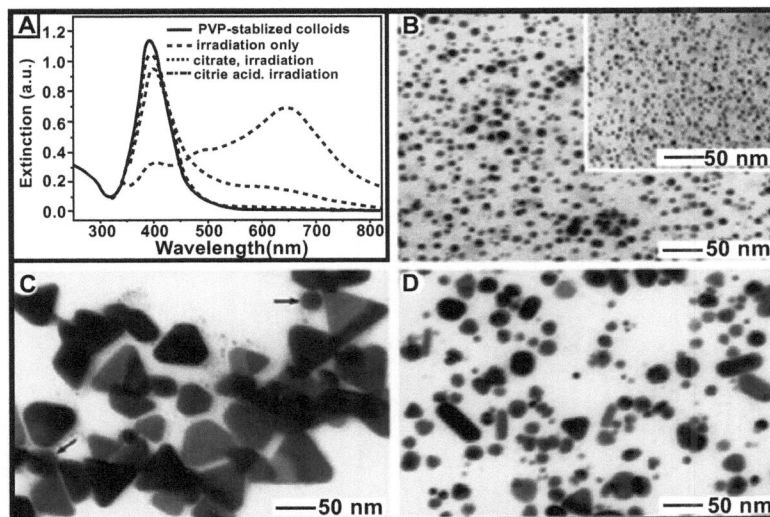

Fig. (4). Silver nanoparticles of different morphologies obtained from PVP and sodium borohydrate solution. Images reprinted with permission from Ref. [32] Copyright © 2003 Wiley-VCH Verlag GmbH & Co. KGaA.

least the minimization of the generated waste and the implementation of sustainable processes through the adoption of fundamental principles. The importance of green methods is not only minimizing the toxic chemicals, the tremendous use of green stabilized nanoparticles can be applied in bio-applications such as MRI and drug delivery systems, due to their biocompatibility [34].

Researchers have begun using biological molecules as templates for generation of inorganic structures and materials. Biological systems form sophisticated mesoscopic and macroscopic structures with tremendous control over the placement of nanoscopic building blocks within extended architectures. Gum acacia polymer promotes the reduction process and act as a good stabilizer for silver nanoparticles, over 5 months [35]. The advantage of this methodology is that it is possible to prepare silver nanoparticles of nearly 2 nm size without using any organic solvent or reducing agent. Raveendran *et al.* [36] have produced silver nanoparticles by green approach. The use of environmentally benign and renewable materials as the respective reducing and protecting agents such as, glucose and starch, as well as a benign solvent medium, offers numerous benefits ranging from environmental safety to ready integration of these nanomaterials. Stable silver nanoparticles have been synthesized by using soluble starch as both reducing and stabilizing agent; this reaction was carried out in an autoclave at 15 psi, 121°C for 5 min [37]. Chen *et al.* [38] have developed silver nanoparticles by using carboxymethyl cellulose sodium salt (CMCS) which can act both as a reducing and stabilizing reagent, while AgNO$_3$ acting as oxidant in the reaction, and no other reagent is needed. The silver nanoparticles prepared in this way are uniform and stable, which can be stored at room temperature for 2 months without any visible change [38]. Wang *et al.* [39] fabricated hollow silver spheres by using mercapto modified starch as template which was later biodegraded by α-amylase. The mercapto groups played an important role in the process of forming starch/silver core–shell structure, which provided nucleation sites for the growth of a silver shell. Removal of the starch core with α-amylase is the key step to form hollow silver spheres, which prevent the shell structures from breaking. Au$^+$/Ag$^-$ bimetallic nanoparticles were designed for nanoscale superstructure fabrication in the presence of strach. The widespread occurrence of these naturally-occurring polysaccharides makes the process amenable to large scale industrial production [40a]. The resulting nanoparticles dispersions exhibited different colors based on the initial compositions, indicating the formation of Au–Ag alloy nanoparticles. The dispersions containing pure silver and gold particles exhibited light yellow and purple colors, respectively. The intermediate compositions resulted in colors that are varying between yellow and red. The corresponding UV-Vis peak shifts are also observed [40b]. Hu *et al.* [41] reported an environmentally benign process for the synthesis of nearly monodisperse silver nanoparticles in large quantities via a microwave-assisted "green" chemistry method in an aqueous system, using basic amino acids, such as L-lysine or L-arginine, as reducing agents and soluble starch as a protecting agent. The presence of amino acids with basicity such as L-lysine or L-arginine, having two amino groups in each molecule, is indispensable for the synthesis of uniform silver nanoparticles. A new method of synthesizing nanoparticles using green materials in a spinning disk reactor (SDR) was also explored recently [42]. The reducing agent and protecting agent were glucose and starch, respectively, either of which are inexpensive and nontoxic materials. Silver particles were prepared by continuously pumping two solutions, which were a mixture of AgNO$_3$ aqueous solution containing protecting agent and another mixture of NaOH aqueous solution containing the reducing agent, into the chamber of the SDR, where a liquid–liquid reaction took place. The reaction time was less than 10 min, which was much shorter than the traditional methods. After washing, redispersed silver particles of 10 nm size were stable for more than 40 days with or without the addition of a dispersing agent. Vigneshwaran et al [43] developed a biosorption mechanism for metal ions by microorganisms which include ion exchange, precipitation and complexation. Reduction and surface accumulation of metals may be occurred by microorganisms protecting themselves from the toxic effects of metallic ions. This study shows the biosorption of silver in the form of nanoparticles by the fungus A. flavus.

2.4. Radiation Method

The homogeneous nanosized silver, copper, gold and nickel particles can be developed in a polymer matrix by using radiation method. Organic monomers and metal compounds were mixed homogeneously at molecular level and the solution is then subjected to irradiation in the field of Co gamma ray (γ-ray) source under normal pressure at room temperature. During this process, the formation of nanocrystalline metal particles as well as polymer formation occurs in one step.

Zhu *et al.* [44] developed γ-radiation method to prepare polyacrylamide–silver nanocomposites at room temperature. In this method, aqueous soluble monomer and metal salt were mixed homogeneously in aqueous solution. When the solution was γ-irradiated, polymerization and reduction took place simultaneously, leading to a homogeneous dispersion of nanocrystalline metal particles in a polymer matrix. The compatibility of monomer and metal salt or aqueous metal salt solution is most important. Further, the application of this method is restricted when the metal salts are insoluble in the monomer or monomer solutions. A new methodology was proposed to obtain polymer–metal nanocomposites by microemulsion in combination with γ- radiation, at room temperature [45]. Au/PAM nanocomposites have been successfully prepared in one step by γ-irradiation in ethanol at room temperature and normal pressure [46]. Ethanol was selected as a solvent since it not only dissolves the reactants but also produces many active products during γ-irradiation, which can induce the polymerization of AM and reduce the metal ions at the same time [46]. Poly(acrylonitrile) (PAN)-silver nanoparticle composites were synthesized using ultraviolet irradiation on a mixture of silver nitrate and acrylonitrile monomer. In addition, a simple way of irradiation technique was employed to synthesize gold nanoparticles at room temperature in an ionic liquid [48]. The quartenary ammonium ionic liquid (QAIL) surrounding the nanoparticle surface can act as an effective stabilizer to gold nanoparticles, and the size of the nanoparticles and uniformity are being affected by the interactions between the QAIL and the clusters during the formation of the nanoparticles. Another radiation method has provided silver nanoparticles with high concentration and with narrow size distribution than those obtained by chemical reduction method, eventhough there was no significant difference between the two strategies in the preparation of gold nanoparticles [49, 50]. γ-Irradiation of 1.0×10^{-3} M AgNO$_3$ solution resulted in nearly 100 times more highly concentrated silver colloids than those by citrate reduction.

3. POLYMER NANOCOMPOSITES (PMNC'S)

In the previous sections, we have discussed the various preparation methods, the reduction and stabilization of bare metal particles, but these bare metal nanoparticle suspensions are normally susceptible to aggregation due to high surface area. In the current section, we demonstrate the nano engineering protocols to improve the properties of polymers by incorporating metal nanoparticles. Consequently, a variety of methods have been developed to modify the metal nanoparticle (MNP) surfaces using organic compounds, in such a way that in polymer networks ranging from small molecules to macromolecules the aggregation of these particles will be prevented. Different types of PMNC's are engineered according to their end use such as drug delivery, antibacterial activity, catalysis, electrical and optical applications. The different types of fabricated PMNCs are microsystems of core shell, nanosystems of core-shell; hollow (shell); core (metal); hydrogels and layer-by-layer etc.

3.1. Micro Bead Systems with Polymer Core and - Metal Shells

Metal nanoparticles when deposited on polymer microbeads look like raspberry and current bun structures. Poly(N-isopropylacrylamide) (PNIPA) microgels can be homogeneously covered with surface-modified gold nanorods [51]. Apart from other interesting effects, they have demonstrated that nanorod optical properties can be used to monitor the thermoresponsive behavior of PNIPA microgel colloids. The surface coverage (the number of rods per microgel sphere) can be controlled simply by varying the microgel to nanorod ratio when mixing the solutions of both (oppositely charged) particle types, as well as through the charge density on the microgel and the nanorod surface. The optical study of the hybrid materials presented here demonstrates that the collapse of the microgel can induce a red shift of the longitudinal plasmon band of the gold nanorods by as much as 28 nm. In addition, changes in the absorbance and the bandwidth also occur, and all these effects are fully reversible when the temperature of the system is lowered again. The reason behind the observed spectral changes lies mainly in the increase of the nanorod density during the shrinking process of the microgel, but the associated changes in the refractive index of the microgel during collapse will also have an influence. In future, they wish to extend this work towards microgels with higher surface charge and intend to explore if charge can affect the red shift.

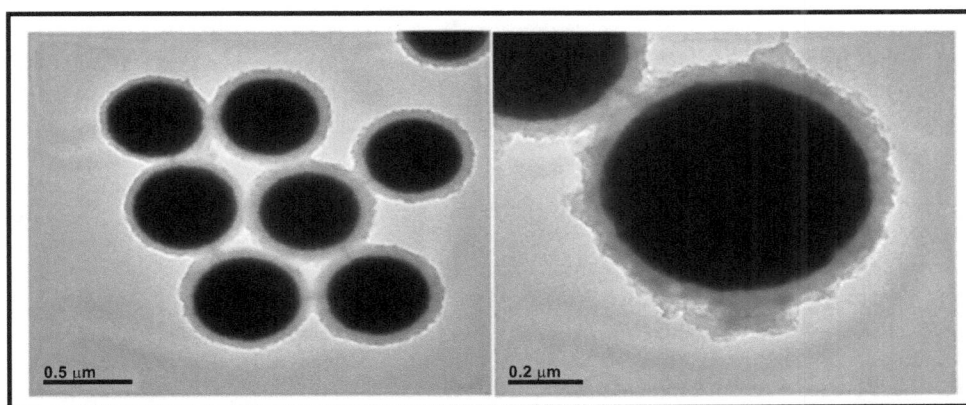

Fig. (5). Ceria nanoparticles coating onto PVP. Images reprinted with permission from Ref. [54] Copyright © 2007 Elsevier.

Moreover, highly charged microgel cores should lead to elevated surface coverages and hence to stronger plasmon-frequency shifts [52]. Kim *et al.* [53] fabricated platinum nanoparticles that are homogeneously immobilized onto the poly(styrene) surface using alcohol-reduction method. Sulfonate groups were employed as a chemical protocol to make a binding between platinum nanoparticle and polystyrene surface. A large number of quasi-spherical platinum nanoparticles with a size of 5 nm in diameter are formed and strongly attached to the PS surface modified with sulfonate groups. Core-shell ceria-PVP nanocomposites were prepared by *in situ* in which mono dispersion of particles and higher percentage of metal content can be achieved as shown in Fig. (**5**) [54].

3.2 Microbeads with Porous Structure

Porous solid materials are of practical significance for useful separation and catalysis, and the main reason is to develop the interconnected pore structure, large surface area, and small pore size [55, 56.]. In order to take advantage of these porous materials, it is a promising approach to introduce functional groups by means of appropriate chemical modification. On the basis of these porous materials, design of nano-sized metal particles into the pores is expected to generate a novel composite system, which helps to define the particle structure and display unique properties [57, 58]. Kim *et al.* [59] synthesized Ag/polymer nanocomposite microspheres by the deposition of Ag in the presence of nitrile (CN)-functional porous polymer microspheres. It was found that the surface characteristics of the supporting microspheres played an important role in determining the degree of deposition of Ag nanoparticles and their distribution in the composite microspheres. This study presents a different colloidal silver (Ag)/polymer system where, Ag nanoparticles are deposited uniformly onto the surface of the functional porous poly (ethylene glycol dimethacrylate-co-acrylonitrile) microbeads. Cereia nanoparticles were developed in microporous beads of poly(4vp-co-dvb) by the suspension polymerization. The polymerization was carried out in the presence of porogens with simultaneous polymerization and oxidation approach [60].

3.3 Polymer Shells with Nanosized Metal Cores

Homogeneously dispersed titanium dioxide (TiO$_2$) / poly (methyl methacrylate) (PMMA) composite microspheres of 1-10 μm size were produced and the interfacial characteristics of TiO$_2$ and PMMA were considered in suspension polymerization. In the electron microscopy, it was observed that TiO$_2$ nanoparticles were embedded homogeneously in the PMMA phase. This study elucidates that the interfacial compatibility between TiO$_2$ and PMMA played a decisive role in producing the composite microspheres structures having inner TiO$_2$ particles in the continuous PMMA phase. This was achieved by treating the surface of the TiO$_2$ particles hydrophobically. The TiO$_2$/PMMA composite microspheres produced in this study showed good ability to protect against UV rays and are therefore of great utility in cosmetic formulations [61a]. Homogeneously zinc oxide (ZnO)-dispersed PMMA composite microspheres were also produced by considering the interfacial characteristics of ZnO and PMMA by in *insitu* suspension polymerization [61b]. These metal particles have improved the conductivity and thermal properties of the polymers, Bhattacharya *et al.* [61c] prepared the zirconium oxide-poly pyrrole nanocomposites, thereby increasing the conductivity of the polypyrrole by 17 times than that of polypyrrole. A simple and easy route for the synthesis of core-shell Ag-PAni nanocomposites was reported recently [62]. No chemical reaction was occurred between the Ag-PAni; and the metallic silver particles were encapsulated in the cores of the growing polymer chains, resulting in the formation of a core-shell Ag-PAni hybrid material. When gamma radiolysis was used for the initial reduction of silver nitrate to silver nanoparticles the later also acted as a mild photocatalyst to oxidize the aniline to polyaniline. Core-shell Ag-PAni nanocomposites are also expected to enhance their application in molecular electronics and other fields. A double-surfactant-layer-assisted polymerization method was proposed to prepare well-controlled core/shell metal oxide/polyaniline (c/s-MO/PANi) nanocomposites. Monodispersed and uniform core/shell nanocomposites including c/s- CuO/PANi, Fe$_2$O$_3$/PANi, −In$_2$O$_3$/PANi and Fe$_2$O$_3$/SiO$_2$/PANi were successfully prepared using this polymerization method [63].

3.4 Metal Nanoparticles Immobilized in Hydrogel Networks

The synthesis of metal nanoparticles within the polymeric network architectures [64,65] and the products of these approaches are of new hybrid or composite systems in chemistry and engineering, It has been reported that colloidal particles present in these gels can be utilized as sensors [66], photonic crystals [67.68], for catalysis [69], switchable electronics [70,71], bio-separation, and drug delivery [66]. In fact, three-dimensional network hydrogels are more suitable as templates for the production of nanoparticles than the conventional non-aqueous or polymeric systems for biomedical applications, considering their exceptional compatibility with biological molecules, cells and tissues. In addition, the main advantages of hydrogels are they provide free space between the networks in the swollen stage Fig. (**6**), which can serve for nucleation, and growth of the nanoparticles. Furthermore, the size and morphology of the nanoparticles can be controlled by functionalization of hydrogels, varying the crosslinking density of networks and modification of the hydrogels [72].

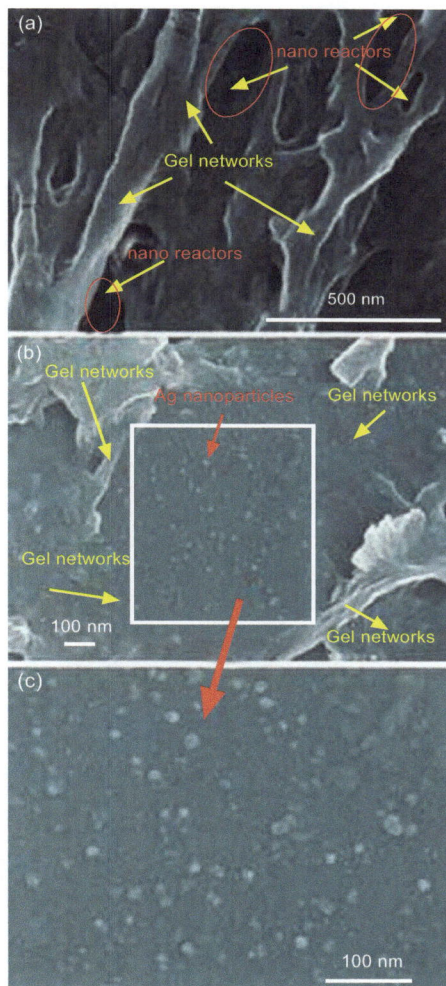

Fig (6). Hydrogel network serve as nanopots for nanoparticle synthesis. Images reprinted with permission from Ref. [72] Copyright © 2007 Elsevier.

Huang *et al.* [73] found that the polymer chains have prohibited the excessive aggregation of the metal atoms and have protective effect on the ultrafine copper particles. Ultrafine copper particles having diameters in the nanometer range have been prepared in a polymer matrix, using ionic aggregates in Cu^{2+}-random poly (itaconic acid-co-acrylic acid) complex as precursor films as well as "microreactors" of controllable size. Controlled sized silver nanoparticles were prepared inside the semi-IPN hydrogel templates containing different carbohydrate polymeric networks and the effect of carbohydrate on the formation of silver nanoparticles with and without reducing agent were also investigated [74]. It was found that some of the silver nanoparticles formed inside the hydrogels without the reducing agent by simple, green and environmental friendly methodology are, useful for antibacterial applications.

3.5. Core–Shell Nanoparticles

Nanostructured core–shell materials containing a metal core, such as gold, surrounded by a polymer layer encapsulating the colloidal metal surface are reported. Polymer or biomolecule coated metals have been employed as building blocks for assembly into new functional materials [75,76]. One-way of tailoring the metal nanoparticles (MNPs) surface is the self-assembly of heteroatom (thiol) functionalized polymers on the surface [77,78]. Size-controlled MNPs can also be prepared with a variety of other methods, such as forming MNPs within polyamidoamine dendrimer nanoreactor [79], and within the micelles of amphiphilic block/star block copolymers[80]. Polymers can also be physically adsorbed on the metal surfaces through specific interactions between the gold and the polymer functional groups, thereby forming shells around the MNPs. For example, poly (vinyl phenol), which serves both as a reducing and stabilizing agent [81] and poly(methylphenylphos-phazene) [82] can be adsorbed to MNPs. However; these polymer shells do not provide a particularly compact core–shell system because of steric restrictions. Core–shell material structures, consisting of an inorganic core and a less compact organic polymer shell, can be prepared with a wide range of functionality, such as block copolymers or end-functionalized polymers on MNPs [83,84]. More compact and well-defined polymer shells around metal nanoparticles such as ceramic micro/nanoparticles were prepared [85-87]. Core shell nanoparticulate systems can also be prepared by grafting technique, in which core–shell materials having a dense polymer shell are prepared by chain initiation from active sites attached to the core surface. Surface-confined living radical [88], living cationic [89] or ring opening metathesis polymerization is normally employed to make such core–shell particles. However, ring-opening polymerization and ionic polymerization are sensitive to moisture and impurities and needs high precaution and therefore difficult to form a shell on MNPs surface that contains traces of water and impurities mixed during the reduction process. In contrast, living free-radical polymerization can tolerate water, air, and some impurities and is applicable to variety of monomers. Therefore, the surface confined living radical polymerization (SCLRP) technique has been widely used as a tool for surface functionalization to generate polymers of controlled molecular weight and well-defined chain-end functionality [90]. With this technique, it is possible to prepare core–shell materials with polymer shells of better uniformity and controlled thickness. Also, the resulting polymer shells can be end-functionalized or block-copolymerized upon the addition of another monomer. A number of research groups, including our own, have employed living radical polymerization such as atom transfer radical polymerization (ATRP) to make core–shell materials consisting of different core materials such as silica, iron oxide, gold, TiO_2, and shells containing poly (sty) poly (benzyl methacrylate) (PBzMA), poly(butyl acrylate), and a block copolymer of styrene and benzyl acrylate. Most work in the area of polymer-coated MNPs synthesis is by SCLRP, however, the reaction has been conducted at ambient or low temperatures. In the case of gold, SCLRP is usually performed from self-assembled monolayers (SAMs) of initiators attached to the surface *via* thiol groups. These reactions are conducted at ambient temperature because of the fragility of the Au-S bond at elevated temperatures in organic solvents. The reported catalyst systems for conducting room-temperature ATRP are limited in number and are not applicable to all vinyl monomers. On the other hand, many catalyst systems are available to carry out ATRP at elevated temperatures for many monomer classes, including acrylates, methacrylates, styrenics, and vinyl acetate. An additional benefit is the ability to prepare thermally stable gold–polymer hybrid core–shell materials. This kind of stable MNPs dispersion can be used for the preparation of well-dispersed polymeric nanocomposites for which high temperature is necessary for molding the raw composite materials into a particular shape. Another method describes the preparation of polymer-coated gold nanoparticles (GNPs) with surface-confined ATRP at elevated temperature [91]. The procedure combines their previously described approaches for making polymer-coated silica microspheres and GNPs. GNPs primed with a thin silica layer containing an initiator were used to initiate living radical polymerization. To solve the problem of thiol desorption, the GNPs were coated with a silica-primer layer through the formation of a 3-mercaptopropyltrimethoxysilane (MPS) SAM on gold, followed by hydrolysis and condensation of the trimethoxysilane groups. This process increased the stability of the monolayer and provided a means of attaching an ATRP initiator to the gold (Au) surface. This modified surface enabled the polymerization with copper (I) chloride (CuCl)/2,2′-bipyridyl as a catalyst system to grow a (PMMA) layer on the gold surface at high temperature. In nanosized hydrogel polymers, photosensitive moieties such as dyes or metal nanoparticles have been incorporated into the hydrogel shell. Jun-Hyun Kim grew a hydrogel layer around a gold nanoparticle core and demonstrated the thermally reversible swelling/deswelling behaviour of the hydrogel coating. This approach enables to produce stable, chemically resistant polymer shells on gold nanoparticle cores [92.93].Uncrosslinked homopolymer poly(*N*-isopropylacrylamide) (NIPAM) has been a particularly well-studied thermosensitive hydrogel. This material, however, enjoys limited applications because of its fixed LCST of 32°C. Colloidal NIPAM hydrogels swell and de-swell at this temperature upon cooling and heating, respectively. Scientists have sought to overcome this limitation by

adding acrylic acid (AAc) in to the NIPAM homopolymer backbone, which can shift the LCST of the copolymer anywhere between 32 to 60°C. This material also exhibits a completely reversible swelling-collapsing behaviour in response to changes in temperature and/or pH. Hydrogel-coated gold nanoparticles were prepared by surfactant-free emulsion polymerization (SFEP) at 70°C in aqueous solution. This research has demonstrated that surfactant-free emulsion polymerization is an efficient method for preparing NIPAM-AAc hydrogel-coated gold nanoparticles in which the thickness of the hydrogel coating varies from 20 to 90 nm. For the gold nanoparticles, the spectra show a maximum absorption band at 530-540 nm arising from the gold Plasmon for the hydrogel-coated gold nanoparticles.

3.6. Hollow Polymer Nanospheres with Movable Metal Cores

Hollow nanospheres functionalized with movable inorganic nanoparticle cores have been explored as novel nanostructures by several groups [94-97]. Core–shell materials can be modified with biomolecules [98] and converted easily into hollow micro/nanospheres by the removal of the core by heating or chemical etching [95,99]. Nanoparticles such as tin, gold, or silica could be incorporated into the interior of hollow nanospheres. Novel properties can be introduced to the hollow nanospheres. Kamata *et al* [95] reported that the incorporation of Au nanoparticle as a movable core into a polymeric hollow sphere could provide an optical probe for monitoring the diffusion of chemical reagent into and out of the shell. Skirtach *et al* [100] showed that, under laser illumination, the capsules containing Ag nanoparticles could be deformed or cut, thus providing a venue for remote release of encapsulated materials. The usual method for the preparation of hollow nanospheres with movable cores is based on a template-assisted approach [95,96] First, the core (e.g., Au nanoparticle) particle is prepared and then it is coated with a polymeric shell. The shell is further functionalized with certain reactive functional groups to grow another polymericlayer. With removal of the middle polymer layer by using a solvent or calcination, hollow spheres with movable cores were formed.

3.7. Layer-by-Layer Assembly

The layer-by-layer (LBL) assembly technique has been proven to be one of the most promising new methods of thin film deposition [101]. The LBL method can be described as alternating deposition of thin films of oppositely charged polyelectrolytes on substrates, and the driving force is usually electrostatic interactions between the two components. Recently, the LBL method has also been successfully applied to thin films of nanoparticles and other inorganic materials [102-104].Its simplicity and universality open a wide range of possible uses for this technique, both in fundamental research and in advanced industrial applications. In LBL assemblies of nanoparticles based on electrostatic interactions, the nanoparticles usually act as one of the two oppositely charged species, which requires that they must have surface charges or be modified by ionic species. Much excellent work on the surface modification of inorganic nanoparticles has been reported, in which thiols, silica, polyelectrolytes, and other materials have been used as the capping agents [105-109].

4. CONCLUSIONS

In the current chapter, we have focused to reveal various methods of synthesis of nanostructures for novel composite materials and nanoparticles. Basic methods of nanoparticle synthesis allowed only regulating size but novel strategies allow to create various structures including core-shell, hollow and layer by layer assemblies and so on. However, there is still need to develop facile and green approaches to generate nanoparticles or nanocomposites for direct bio-medical applications

5. REFERENCES

[1]　Wang Z, Pinnavaia TJ. Hybrid Organic–Inorganic Nanocomposites: Exfoliation of Magadiite Nanolayers in an Elastomeric Epoxy Polymer. Chem Mater. 1998; 10(7): 1820-6.

[2]　Carotenuto GC, Her YS, Matijevic E. Preparation and Characterization of Nanocomposite Thin Films for Optical Devices. Ind Eng Chem Res. 1996; 35(9): 2929-32.

[3]　LiraCantu M., Gomez Romero P. Chem Mater. 1998; 10: 698.

[4]　Tunney J J, Detellier C. . Aluminosilicate Nanocomposite Materials. Poly(ethylene glycol)–Kaolinite Intercalate Chem Mater 1996; 8(4): 927-35.

[5]　Wang Y, Herron N. Photoconductivity of CdS nanocluster-doped polymers Chem Phys Lett. 1992; 200(1-2),71-5.

[6]　Giannelis E. P. Adv Mater. 1996; 8: 29.

[7]　Herron N, Thorn D. L. Nanoparticles: Uses and Relationships to Molecular Cluster Compounds Adv Mater. 1998; 10: 1173.

[8]　Ouahab L. Organic/Inorganic Supramolecular Assemblies and Synergy between Physical Properties Chem Mater. 1997; 9(9): 1909-26.

[9a]　Zavyalov SA, Pivkina AN, Schoonman J. Formation and characterization of metal-polymer nanostructured composites Solid State Ionics 2002; 147: 415-9.

[9b]　Smith TW, Wychick D. J Phys Chem. 1980; 84: 621.

[10]　Hess PH, Parker Jr. PH. Polymers for Stabilization of Colloidal Cobalt Particles J Appl Polym Sci. 1966; 10: 1915-27.

[11]　Pomagailo AD. Polymer-immobilised nanoscale and cluster metal particles. Russ Chem Rev 1997; 66: 679.

[12]　Chan YNC, Schrock RP, Cohen RE. Chem. Mater. 1992; 4: 24.

[13]　Chan YNC, Crig GSW, Schrock RP, Cohen RE. Synthesis of palladium and platinum nanoclusters within microphase-separated diblock copolymers Chem. Mater.1992; 4: 885.

[14]　Sohn BH. Cohen, RE, Electrical Properties of Block Copolymers Containing Silver Nanoclusters Within Oriented Lamellar Microdomains. J. Appl. Polym. Sci. 1997; 65: 723-9.

[15]　Hashimoto T, Harada M, Sakamoto N. Incorporation of Metal Nanoparticles into Block Copolymer Nanodomains via in-Situ Reduction of Metal Ions in Microdomain Space Macromolecules 1999; 32: 6867-70.

[16]　Lee JY, Liao Y, Nagahata R , Horiuchi S Effect of metal nanoparticles on thermal stabilization of polymer/metal nanocomposites prepared by a one-step dry process Polymer 2006; 47: 7970-9.

[17]　S. Horiuchi, Muhammad I. Sarwar, Yukimichi NakaoNanoscale Assembly of Metal Clusters in Block Copolymer Films with Vapor of a Metal-Acetylacetonato Complex Using a Dry Process Adv. Mater. 2000; 12(20): 1507-11.

[18]　Tannenbaum R,Goldberg EP, Flenniken CL, Decomposition of iron carbonyls in solid polymer matrices.[in] Sheats JE, Carraher CE Jr,Ch.U. Pittman Jr (Eds.), Metal-containing polymeric system, Plenum Press, New York, 1985; 303.

[19]　Abes JI, Cohen RE, Ross CA, Mat. Sci. Eng. C, 2003; 23: 641.

[20]　Clay RT, Cohen, Synthesis of metal nanoclusters within microphase-separated diblock copolymers: sodium carboxylate vs carboxylic acid functionalization. Supramol. Sci, 1998; 5: 41-48.

[21]　Wojciechowska DW, Jeszka JK, Uznanski P, Amiens C, Chaudret B, Lecante P Synthesis of gold nanoparticles in solid state by thermal decomposition of an organometallic precursor Mate Sci-Poland, 2004; 22(4): 407-13.

[22]　Patakfalvi R,. Vira´nyi Z, De´ka´ny I. Kinetics of silver nanoparticle growth in aqueous polymer solutions Colloid Polym Sci 2004; 283: 299-305.

[23]　Ohde H, Hunt F, Chien MW Synthesis of Silver and Copper Nanoparticles in a Water-in-Supercritical-Carbon Dioxide Microemulsion Chem. Mater. 2001, 13, 4130.

[24]　Olenin AY, YA Krutyakov, Kudrinskii AA, Lisichkin GV. Formation of Surface Layers on Silver Nanoparticles in Aqueous and Water–Organic Media Colloid J 2008; 70(1): 71-6.

[25]　Khanna PK , N Singh, S Charan, Subbarao VVVS, Gokhale R , Mulik UP. Synthesis and characterization of Ag/PVA nanocomposite by chemical reduction method .Mater Chem Phys 2005; 93: 117-21.

[26]　Songping, W, Shuyuan M. Preparation of ultrafine silver powder using ascorbic acid as reducing agent and its application in MLCI Mater Chem Phy 2005; 89: 423-7.

[27]　M Habib Ullah, Kim I, Chang-Sik Ha Preparation and optical properties of colloidal silver nanoparticles at a high Ag+ concentration Mate Lett 2006; 60: 1496-501.

[28]　Ullah, Md. Habib; Kim, I; Ha, Chang-Sik J Nano sci Nanotech 2006; 6(8): 777-82.

[29]　Pastoriza-Santos I, Luis M Liz-Marza´n Synthesis of Silver Nanoprisms in DMF Nano Lett 2002; 2(8): 903-5.

[30]　Nersisyan HH, Lee JH, Son HT, Won CW, Maeng DY. A new and effective chemical reduction method for preparation of nanosized silver powder and colloid dispersion Mater Resear Bulle 2003; 38: 949.

[31]　Gao Y, Jiang P, Songa L, Wanga JX, Liua LF, Liua DF, Xianga YJ, Zhanga ZX, Zhaoa XW, Doua XY, Luoa SD, Zhoua WY, Xiea SS, Studies on silver nanodecahedrons synthesized by PVP-assisted N,N-dimethylformamide (DMF) reduction. J Crystal Growth 2006; 289: 376-80.

[32]　Sun Y, Xia Y triangular nanoplates of silver: synthesis characterization, and use as sacrificial templates for generating triangular nanorings of gold. Adv Mater 2003; 15: 695.

[33]　He R, Qian X, Yin J, Zhu Z. Formation of silver dendrites under microwave irradiation Chemical Physics Letters 2003; 369: 454-8.

[34]　Dhar S, Maheswara Reddy E, Shiras A, Pokharkar V, Prasa BLV Natural Gum Reduced/Stabilized Gold Nanoparticles for Drug Delivery Formulations Chem. Eur. J. 2008; 14: 1024-30.

[35] Mohan YM, Raju KM, Sambasivudu K, Singh S, Sreedhar B. Preparation of Acacia-Stabilized Silver Nanoparticles: A Green Approach J Appl Polym. Sci, 2007; 106: 3375-81.

[36] Raveendran P, Fu J,. Wallen SLCompletely "Green" Synthesis and Stabilization of Metal Nanoparticles. J. Am. Chem. Soci. 2003; 125(46): 13940-1.

[37] Vigneshwaran N, Nachane RP, Balasubramanya RH, Varadarajan PV A novel one-pot 'green' synthesis of stable silver nanoparticles using soluble starch Carbohydrate Research 2006; 341: 2012-8.

[38] Chen J, Wang J, Zhang X, Jin Y Microwave-assisted green synthesis of silver nanoparticles by carboxymethyl cellulose sodium and silver nitrate. Mater Chem Phy 2008; 108: 421-4.

[39] Wang L, Chen D A facile method for the preparation of hollow silver spheres Mater Lett 2007; 61: 2113-6.

[40a] Haizhen Huang and Xiurong Yang Synthesis of polysaccharide-stabilized gold and silver nanoparticles: a green method Carbohydrate Research 2004; 339: 2627-31.

[40b] Raveendran P, Scott Fu J, Wallen L. A simple and "green" method for the synthesis of -Au, Ag, and Au–Ag alloy Nanoparticles Green Chem., 2006; 8: 34-38.

[41] Hu B, Wang SB, Wang K, Zhang M Yu SH, Microwave-Assisted Rapid Facile "Green" Synthesis of Uniform Silver Nanoparticles: Self-Assembly into Multilayered Films and Their Optical Properties. J. Phys. Chem. C 2008; 11(30)2: 11169-74.

[42] Clifford Y, Wang YH, Liu HS A Green Process for Preparing Silver Nanoparticles Using Spinning Disk Reactor AIChE J 2008; 54(2): 445-52.

[43] Vigneshwaran N, Ashtaputre NM, Varadarajan PV, Nachane RP, Paralikar KM. Balasubramanya RH. Biological synthesis of silver nanoparticles using the fungus Aspergillus flavus Mater Lett 2007; 61: 1413-8.

[44] Zhu Y, Qian Y, Li X , Zhang M g-Radiation synthesis and characterization of polyacrylamide–silver nanocomposites Chem. Commun., 1997; 82: 1081-2.

[45] Yin Y, Xu X, Chuanjun Xia, Ge X, Zhang ZSynthesis and characterization of poly(butyl acrylate-co-styrene)–silver nanocomposites by g radiation in W/O microemulsions Chem. Commun., 1998; 941-2.

[46] Ni Y, Ge X, Zhang Z, Ye Q In situ single-step synthesis of gold/polyacrylamide nanocomposites in an ethanol system Mate Lett 2002; 55: 171.

[47] Zhang L, Wang S, Chen W, Lei a convenient route to polyacrylonitrile/silver nanoparticle compositeby simultaneous polymerization and reduction approach. Polymer 2001; 42: 8315-18.

[48] Chen S, Guozhong Wu YL Stabilized and size-tunable gold nanoparticles formed in a quaternary ammonium-based room-temperature ionic liquid under γ-irradiation Nanotechnology 2005; 16: 2360-4.

[49] Li T, Park HG, Choi S-H. Irradiation-induced preparation of Ag and Au nanoparticles and their characterizations Mater Chem Phy 2007; 105: 325-30.

[50] Hyeon M, Shin S, Yang HJ, Kim SB, Lee MSMechanism of growth of colloidal silver nanoparticles stabilized by polyvinyl pyrrolidone in γ -irradiated silver nitrate solution J Colloid Inter Sci 2004; 274: 89-94.

[51] Karg M, Pastoriza-Santos I, P_rez-Juste J, Hellweg T, Liz-Marz LM. Nanorod-coated PNIPAMMicrogels: Thermoresponsive Optical Properties small 2007; 3(7): 1222-9.

[52] Gorelikov I, Field L M Kumacheva E. Hybrid Microgels Photoresponsive in the Near-Infrared Spectral Range J. Am. Chem. Soc. 2004; 126(49): 15938-39.

[53] Kim DW, Lee JM, Chul Oh, Kim DS, Oh SG. A novel preparation route for platinum–polystyrene heterogeneous nanocomposite particles using alcohol-reduction method. J Colloid and Interface Sci 2006; 297: 365.

[54] Sivudu KS, Shailaja D One-step synthesis and characterization of poly(4vp-co-dvb)/ceria nanocomposite by simultaneous polymerization–oxidation approach Materials Letters 2007; 61: 2167-9.

[55] Bu H, Rong J, Yang Z. Template Synthesis of Polyacrylonitrile-Based Ordered Macroporous Materials and Their Derivatives Macromol Rapid Commun. 2002; 23(8): 460- 4.

[56] Jiang P, Hwang KS, Mittleman DM, Bertone JF, Colvin VL. Template-Directed Preparation of Macroporous Polymers with Oriented and Crystalline Arrays of Voids J Am Chem Soc. 1999; 121(50): 11630-7.

[57] Dokoutchaev A, James JT, Koene SC, Pathak S, Surya GK, Thompson ME. Colloidal Metal Deposition onto Functionalized Polystyrene Microspheres Chem Mater. 1999; 11(9): 2389-99.

[58] Jun JB, Kim JW, Lee JW, Suh KD. Spherical Polarization Body: Synthesis of Monodisperse Micron-Sized Polyaniline Composite Particles Macromol Rapid Commun. 2001; 22(12): 937-40.

[59] Kim JW, Lee JE., Kim SJ., Lee JS, Ryu JH, Kima J, Hana SH, Ih-Seop Chang, Suh KD. Synthesis of silver/polymer colloidal composites from surface-functional porous polymer microspheres. Polymer. 2004; 45: 4741-7.

[60] Sivudu KS, Reddy YB. Yadav JS, Sabitha G. Shailaja D. Ceria-Supported Vinylpyridine Polymers: Synthesis, Characterization and Application in Catalysis Inter J Polym Mater , 2008; 57: 891-03.

[61a] JW Kim, JW Shim, J-H Bae, SH Han, HK Kim, IS Chang,HH Kang, KD Suh. Titanium dioxide/poly(methyl methacrylate) composite microspheres prepared by in situ suspension polymerization and their ability to protect against UV rays Colloid Polym Sci 2002; 280: 584-8.

[61b] JW Shima, JW Kim, SH Han, IS Chang,HK Kim, HH Kang, OS Lee, KDo Suh Zinc oxide/polymethylmethacrylate composite microspheres by in situ suspension polymerization and their morphological study Colloids and Surfaces A: Physicochemical and Engineering Aspects 2002; 207: 105-111.

[61c] Bhattacharya A, Ganguly KM , De A, Sarkar S. A new conducting nanocomposite-ppy-zirconium (iv) oxide Mate Rese Bull. 1996; 31(5): 527-30.

[62] Rezaulkarim M, Taeklim K, Lee C J, Huiyan MTI, Kim HJ, Park Mu Samg Lee LS. J Polym Sci: Part A: 2007; 45: 5741.

[63] Zhu CL, Chou SW, He SF, Liao WN , Chen CC. Synthesis of core/shell metal oxide/polyaniline nanocomposites and hollow polyaniline capsules Nanotechnology 2007; 18: 275604-10.

[64] Aymonier C, Schlotterbeck U, Antonietti L, Zacharias P, Thomann R, Tiller JC, Mecking S. Hybrids of silver nanoparticles with amphiphilic hyperbranched macromolecules exhibiting antimicrobial properties Chem Commun. 2002; 3018-19.

[65] Ho CH, Tobis J, Sprich C, Tomann R, Tiller JC. Nanoseparated Polymeric Networks with Multiple Antimicrobial Properties. Adv. Mater. 2004; 16(12): 957-61.

[66] Li YY, Cunin F, Link JR, Gao T, Betts RE, Reiver SH, Chin V, Bhatia SN, Sailor MJ Polymer Replicas of Photonic Porous Silicon for Sensing and Drug Delivery Applications Science. 2003; 299: 2045.

[67] Biffis A, Orlandi N, Corain B. additive driven phase selective chemistry in block copolymer thin films: the coveregence of top down bottom-up approaches. Adv. Mater. 2003; 15(18): 1551-5.

[68] Xu S, Zhang J, Paquet C, Lin Y, Kumacheva E. from hybrid microgels to photonic crystals Adv Funct Mater 2003; 13(6): 468-72.

[69] Aymonier C, Bortzmeyer D, Thomann R, Mu¨lhaupt R. Poly(Methyl methacrylate)/Palladium Nanocomposites: Synthesis and Characterization of the Morphological, Thermomechanical, and Thermal Properties. Chem. Mater. 2003; 15(25): 4874-78.

[70] Holland B. T, Blanford CF, Stein A. Science. Ordered Three- Dimensional Arrays of Spheroidal Synthesis of Macroporous Minerals with Highly voids. 1998; 281: 538-40.

[71] Kazimierska EA, Ciszkowska M. Thermoresponsive Poly-N-isopropylacrylamide Gels Modified with Colloidal Gold Nanoparticles for Electroanalytical Applications. Preparation and Characterization. Electroanalysis. 2005; 17(15-16): 1384-95.

[72] YM Mohan, TP kumar, K Lee, KE. Geckeler Hydrogel networks as nanoreactors: A novel approach to silver nanoparticles for antibacterial applications Polymer. 2007, 48, 158-64.

[73] Huang C. Chen L, Yang C Study of Cu2+-poly(itaconic acid-*co*-acrylic acid) complex and copper-polymer nanocomposites. Polym Bull. 1998; 41: 585-92.

[74] Vimala K, Sivudu KS , Mohan YM, Sreedhar B, Raju KM. Controlled silver nanoparticles synthesis in semi-hydrogel networks of poly(acrylamide) and carbohydrates: A rational methodology for antibacterial application. Carbohydrate Polym. 2009; 75: 463–71.

[75] Oal AK, Ilhan F, DeRouchey JE, Thurn-Albrecht T, Russell TP, Rotello VM. Self-assembly of nanoparticles into structured spherical and network aggregates. Nature. 2000; 404, 746-8.

[76] Simard J, Riggs C, Boal AK, Rotello VM. Formation and pH-controlled assembly of amphiphilic gold nanoparticles. Chem Commun. 2000; 1943-4.

[77] Chechik V, Crooks RM. Monolayers of Thiol-Terminated Dendrimers on the Surface of Planar and Colloidal Gold. Langmuir 1999; 15(19): 6364-69.

[78] Wuelfing WP, Gross SM, Miles DT, Murray RW. Nanometer Gold Clusters Protected by Surface-Bound Monolayers of Thiolated Poly(ethylene glycol) Polymer Electrolyte. J Am Chem Soc. 1998; 120(48), 12696-97.

[79] Groehn F, Auer BJ, Akpalu YA, Jackson CL, Amis EJ. Dendrimer Templates for the Formation of Gold Nanoclusters Macromolecules. 2000; 33(16): 6042-50.

[80] Spatz JP, Mossmer S, Hartmann C, Moller M, Herzog T, Krieger M, Boyen HG, Ziemann P, Kabius B. Ordered Deposition of Inorganic Clusters from Micellar Block Copolymer Films Langmuir. 2000; 16(2): 407-15.

[81] Bhattacharjee RR, Chakraborty M, Mandal TK. Synthesis of size-tunable gold nanoparticles by poly(vinylphenol) and electrostatic multil J Nanosci Nanotech. 2004; 4: 844-8.

[82] Walker CH, St. John JV, Wisian-Neilson P. Synthesis and Size Control of Gold Nanoparticles Stabilized by poly(methylphenylphosphazene) J Am Chem Soc. 2001; 123(16): 3846-7.

[83] Song Q, Ai X, Preparation of Gold/triblock Copolymer Composite Nanoparticles. J Nanopart Res. 2000; 2: 381-5.

[84] Marinakos SM, Brousseau LC, Jones A, Feldheim DL. Template Synthesis of One- Dimensional Au, Au-Poly(pyrrole) and Poly(pyrrole) Nanoparticle Arrays. Chem Mater. 1998; 10: 1214-9.

[85] Huang X, Wirth M. J. Anal Chem. 1997; 69: 4577. Huang X, Wirth MJ. Surface Initiation of Living Radical Polymerization for Growth of Tethered Chains of Low Polydispersity Macromolecules. 1999; 32(5): 1694-6.

[86] Von Werne T, Patten TE. Preparation of Structurally Well-Defined polymer–Nanoparticle Hybrids with Controlled/Living Radical Polymerizations J Am Chem Soc. 1999; 121(32): 7409-10.

[87] Von Werne T, Patten TE. Atom Transfer Radical Polymerization from Nanoparticles: A Tool for the Preparation of Well-Defined Hybrid Nanostructures and for Understanding the Chemistry of Controlled/"Living" Radical Polymerizations from Surfaces. J Am Chem Soc 2001; 123(31): 7497-505.

[88] Mandal TK, Fleming MS, Walt DR. Preparation of Polymer Coated Gold Nanoparticles by Surface-Confined Living Radical Polymerization at Ambient Temperature. Nano Lett. 2002; 2: 3-7.

[89] Jordan R, Ulman A. Surface Initiated Living Cationic Polymerization of 2-Oxazolines. J Am Chem Soc. 1998; 120: 243.

[90] Bontempo D, Tirelli N, Masci G, Crescenz V, Hubbell JA. Thick Coating and Functionalization of Organic Surfaces via ATRP in Water Macromol Rapid Commun. 2002; 23(7)417-22.

[91] Kotal A, Mandal TK, Walt DR Synthesis of Gold–Poly(methyl methacrylate) Core–Shell Nanoparticles by Surface-Confined Atom Transfer Radical Polymerization at Elevated Temperature J Polym Sci Part A: 2005; 43: 3631-42.

[92] Kim JH, Lee TR Discrete Thermally Responsive Hydrogel-Coated Gold Nanoparticles for Use as Drug-Delivery Vehicles. Drug development research 2006; 67: 61-69.

[93] Kim JH, Lee TR. Thermo- and pH-responsive hydrogel-coated gold nanoparticles. Chem. Mater. 2004, 16, 3647-51.

[94] Lee KT, Jung YS, Oh SM. Synthesis of Tin-Encapsulated Spherical Hollow Carbon for Anode Material in Lithium Secondary Batteries J. Am. Chem. Soc. 2003; 125(19): 5652-3.

[95] Kamata K, Lu Y, Xia Y. Synthesis and Characterization of Monodispersed Core−Shell Spherical Colloids with Movable Cores J. Am. Chem. Soc. 2003; 1259(9)2384-5.

[96] Zhang K, Zhang X, Chen H, Chen X, Zheng L, Zhang J. Yang B. Hollow Titania Spheres with Movable Silica Spheres Inside Langmuir. 2004; 20(26): 11312-4.

[97] Kim M, Sohn K, Na HB, Hyeon T. Synthesis of Nanorattles Composed of Gold Nanoparticles Encapsulated in Mesoporous Carbon and Polymer Shells Nano Lett. 2002; 2(12): 1383-7.

[98] Quaroni L, Chumanov G. Preparation of Polymer-Coated Functionalized Silver Nanoparticles J Am Chem Soc. 1999; 121(45): 10642-3.

[99] Bhattacharjee RR., Chakraborty M, Mandal TK. Synthesis of dendrimer-stabilized gold- polypyrole core-shell nanoparticles. J Nanosci Nanotechnol. 2003; 3(6): 487-91.

[100] Skirtach AG., Antipov AA, Shchukin DG, Sukhorukov GB. Remote Activation of Capsules Containing Ag Nanoparticles and IR Dye by Laser Light Langmuir 2004; 20(17): 6988-92.

[101] Decher GD. Fuzzy Nanoassemblies: Toward Layered Polymeric Multicomposites Science. 1997; 277: 1232-7.

[102] Rogach AL, Koktysh DS, Arrison M, Kotov NA. Layer-by-Layer Assembled Films of HgTe Nanocrystals with Strong Infrared Emission Chem. Mater. 2000; 12(6): 1526-8.

[103] Ostrander JW, Mamedov AA, Kotov NA. Two Modes of Linear Layer-by-Layer Growth of Nanoparticle−Polyelectrolyte Multilayers and Different Interactions in the Layer-by-layer Deposition J. Am. Chem. Soc. 2001; 123(6): 1101-10.

[104] Pastoriza-Santos I, Koktysh DS, Mamedov AA, Giersig M, Kotov NA, Liz-Marzan L M. One-Pot Synthesis of Ag@TiO2 Core−Shell Nanoparticles and Their Layer-by-Layer Assembly Langmuir. 2000; 16(6): 2731-5.

[105] Kotov NA, Dekan I, Fendler JH. Layer-by-Layer Self-Assembly of Polyelectrolyte - Semiconductor Nanoparticle Composite Films. J. Phys. Chem. 1995; 99: 13065-9.

[106] Bright RM, Musick MD. Natan MJ. Preparation and Characterization of Ag Colloid Monolayers Langmuir. 1998; 14(20): 5695-701.

[107] Cassagneau T, Fendler JH. Preparation and Layer-by-Layer Self-Assembly of Silver Nanoparticles Capped by Graphite Oxide Nanosheets J. Phys. Chem. B, 1999; 103(11): 1789-93.

[108] Lvov YM, Rusling JF, Thomsen DL, Papadimitrakopoulos F, Kawakami T, Kunitake T. High-speed multilayer film assembly by alternate adsorption of silica nanoparticles and linear polycation Chem. Commun. 1998; 1229-30.

[109] Mamedov A, Ostrander JW, Aliev F, Kotov NA. Stratified Assemblies of Magnetite Nanoparticles and Montmorillonite Prepared by the Layer-by-Layer Assembly Langmuir. 2000; 16(8): 3941-9.

CHAPTER 7

SYNTHESIS OF METAL NANOPARTICLES USING HYDROGEL NETWORKS

Varsha Thomas[1], Y. Murali Mohan[2*], Manjula Bajpai[1] and S.K. Bajpai[1*]

[1]*Department of Chemistry, Polymer Research Laboratory, Government Model Science College, Jabalpur, MP 482001, India;* [2]*Cancer Biology Research Center, Sanford Research/ USD, Sioux Falls, SD-57105*

Address correspondence to: Murali Mohan, Cancer Biology Research Center, Sanford Research/ USD, Sioux Falls, SD-57105; E-mail: mohanaym@yahoo.co.in or yallapum@sanfordhealth.org; S.K. Bajpai, 1Department of Chemistry, Polymer Research Laboratory, Government Model Science College, Jabalpur, MP 482001, India; E-mail: mnlbpi@rediffmail.com

Abstract: Hydrogels containing silver nanoparticles have a variety of biomedical applications because of the excellent compatibility of hydrogels with biological molecules, tissues, cells, and exceptional antibacterial properties of silver nanoparticles. Here we report the design and development of various hydrogel networks that can be employed as matrices to grow and form metal nanoparticles. This behavior is achieved because of their enormous hydrophilic groups and free space between gel networks that extend not only stabilization but also participate in reduction process. This chapter provides an in-depth review of various hydrogel systems that can be employed for nanoparticles synthesis.

Key words: Hydrogel, microgel, silver nanoparticles, antibacterial applications.

1. INTRODUCTION

Nanoparticles (NP) are composed of several tens or hundreds of atoms or molecules; and can have defined sizes and morphologies (amorphous, crystalline, spherical, rod, needles, etc.). They are larger than atoms and molecules but are smaller compared to bulk solids. The nanoparticles obey neither laws of absolute quantum chemistry nor laws of classical physics; but overall, they exhibit their physico-chemical properties that differ markedly from those expected. For example, substances that are insulators in bulk form might become semiconductors when reduced to the nanoscale. Similarly, copper nanoparticles smaller than 50 nm are considered super hard materials that do not exhibit in bulk copper. The properties of materials change as their size approaches the nanoscale and the percentage of atoms at the surface of a nanomaterial becomes significant. The interesting and unexpected properties of nanoparticles are partly due to the aspects of the surface of the material dominating from the bulk properties. Nanoparticles include metal nanoparticles, carbon nanotubes, semiconductor quantum dots and other particles produced from huge variety of substances.

Few nanoparticle formulations are readily available commercially in the form of dry powders, liquid/semi-liquid dispersions or fluid gels. It is always advisable to use chemical additives (surfactants or dispersing agents) to obtain a uniform and stable dispersion of nanoparticles formulations. Nanoformulations are being used to fabricate coatings, components or devices, and many biological applications. Industrial scale production of some types of nanoparticulate materials like carbon black, polymer dispersions or micronized drugs (drug loaded or drug entrapped nanoparticles) has been established for a long time.

Nanoparticles often have physical and chemical properties that are very different from the parent bulk material. The properties of nanoparticles depend on their size, shape, surface characteristics and internal structure. Metal and metal oxide nanoparticles are extensively used in various fields. Nanoparticles labeled microchips and microchips prepared by NPs are used for detecting chemical, biological and radiological agents. In these chips, NPs are first laid followed by a layer of a special polymer, and then a layer of receptor molecules [1]. Lu *et al.* [2] demonstrated that the gold nanoparticles (AuNP)s, silver nanoparticles (AgNP)s, and gold nanoshells can be used as optical volatile organic compounds (VOCs) sensors. The surface area of these nanomaterials is sufficiently high for quantitative adsorption of VOCs. The response of localized surface plasmon resonance (LSPR) spectra of gold and silver nanoparticles was sensitive to changes in extinction, while gold nanoshells exhibited red-shifts in wavelength when exposed to organic vapors. Both Ag and Au NPs exhibit very high conductivity ($\sim10^4$-10^5 Scm^{-1}). AuNPs are highly applicable in various electrical devices [3, 4] but the high cost of gold negates their usage. AgNPs are equally efficient to prepare electrical devices like 3 D micro-electromechanical systems and electrical circuitry [5, 6]. A novel, water-dispersible oleic acid (OA)-Pluronic-coated iron oxide magnetic nanoparticle formulation was synthesized for systemic administration of anti-cancer drugs while simultaneously allowing magnetic targeting and/or imaging [7]. In addition, AuNPs are also used in sensors for both chemical and biological warefares [8].

Recent literature suggests the use of metal nanoparticles (Ag and Au) in biomedical applications [9]. For these applications, generation of novel and green approaches is highly essential. Therefore, in this book chapter we focus on general chemistry involved in the synthesis of metal nanoparticles and how hydrogel networks serve as nanoreactors for metal nanoparticles. These hydrogel-metal nanocomposites are highly bio-compatible and can be used directly for medical applications.

2. SYNTHESIS OF METAL NANOPARTICLES

Physical methods, such as, mechanical smashing, solid-phase reactions, freeze-drying, spray-drying and precipitation, are commonly used to prepare nanoparticles. However, physical methods consume lot of energy to maintain high pressure and temperatures. For example, silver nanoclusters were synthesized within the microdomains from the silver-complexes by heating upto 120°C for 24 h under vacuum [10]. Such conditions are hard to maintain for long time and these methods are not economically acceptable.

Many organic and inorganic compounds are being used for the synthesis of NPs. The organic reducing agents, such as hydrogen [11], sodium borohydride [12], hydrazine [13], hydrogen peroxide [14], hydroxylamine [15], formaldehyde and its derivatives [16], ascorbic acid [17], ethylenediaminetetraacetic acid (EDTA) [18], potassium bitartrate [19], ethanol [10], ethyl glycol [21], organometallic precursors [22-24] and dimethylene sulphoxide (DMSO) [25] are widely used for the synthesis of NPs. Most of these reducing agents are toxic and can restrict the nanoparticles application for certain fields but not medical purpose. For example, reduction of silver ions by sodium borohydride causes adsorption of borate molecules on the surfaces of the NPs, thus making them unsuitable for the SERS study of active site of mammalian liver microsomal cytochrome P-450 enzyme [26]. Thus, some natural compounds like sodium citrate or reducing sugars are preferably used in the case of chemical synthesis of NPs. The rate of reducing action of these compounds is low and therefore the reduction reaction takes longer times or

the reactions need to be conducted at higher temperatures. However, these reducing agents are mild, inexpensive, and nontoxic. A few categories of these compounds simultaneously play a dual role of protective and reducing agent. The silver particles prepared by citrate reduction result in the formation of large diameters (50–100 nm) with wide range distributions in size and shape. Citrates can serve not only the dual role as reductant and stabilizer but also play the role of a template [27]. The reaction mechanism of formation of silver nanoparticles with trisodium citrate can be shown as follows:-

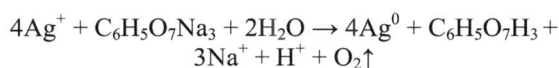

$$4Ag^+ + C_6H_5O_7Na_3 + 2H_2O \rightarrow 4Ag^0 + C_6H_5O_7H_3 + 3Na^+ + H^+ + O_2\uparrow$$

Reduction of metals, as seen above, requires high temperature, reducing agents, tedious conditions and a lot of time. But a well-balanced fabrication is necessary to control the cost, minimize the time duration and protect the environment. The radiation method can fulfil some of these criteria. The advantage of gamma irradiation method is that high number of reducing radicals could be generated for the synthesis of metallic nanoparticles without the formation of any byproducts. The primary radicals and molecules produced in water upon gamma irradiation are e^-_{aq} , $OH^{\cdot}$, $H^{\cdot}$,H_2 and H_2O_2 [28]. Microwave radiation is also used for nanoparticles synthesis at a faster rate. This rapid microwave heating provides uniform nucleation and growth conditions, leading to homogeneous nanomaterial with smaller size and morphology. Power dissipation is fairly uniform throughout with "deep" inside-out heating of the polar solvents, which leads to nanocrystal formation.

The above mentioned physical and chemical methods are hard to be maintained for a long time and are not economically acceptable. At this junction, green methods are paid much attention [29] in which polysaccharides, β-D-glucose or heparin act as reducing agents to syntheses NPs [29, 30]. In addition, biological reduction process allows obtaining highly dispersed NP formulations because of sufficient material sources and mild reaction conditions. Owing to the environment friendly nature of biosynthetic process, microbes like bacteria, fungi, and algae, are able to produce NPs [31-33]. Some reports suggest that fungi have some advantages over bacteria in the nanoparticles formation. This is because the filamentous fungi have high tolerance towards metals, high wall-binding capacity as well as intracellular metal uptake capabilities which permit to culture on large scales.

Besides all these methods, a versatile preparation technique is under implication to produce NPs i.e., micro emulsion. This process enables the control of particle properties, such as size, geometry, morphology, homogeneity and surface area [34-35]. One of the major advantages is that it allows the preparation of metal-based catalysts displaying high surface area and high catalytic activity in the nanosize range. Although there are many surfactants available for forming microemulsion to prepare nanomaterials, sodium bis (2-ethylhexyl) sulfosuccinate (AOT) is the most common surfactant used to form reverse micelles [36-37]. In this system, water is readily solubilized in the polar core, forming "water-pools" characterized by the molar ratio of water to surfactant (W). The "water-pools" are spherical and monodisperse aggregates, and the water-pool radius, Rw, is linearly dependent on the water content. The formed particles have higher stability, smaller particle size and good monodispersity [38-39]. The size and shape of the final nanoparticles are also controlled by the droplet's size and shape [40]. However, microemulsion method also has some drawbacks. For example, the large amounts of surfactant and organic solvent, which are difficult to be separated and removed from the final products, are added to the system and thus it is expensive to fabricate nanoparticles by this method. Furthermore, microemulsion method described previously produced stable silver dispersions only at a relatively low concentration and employed high deleterious organic solvents.

3. GEL NETWORKS FOR SYNTHESIS OF NANOPARTICLES

As the field of nanotechnology is advancing and most sought after, biomedical applications require stable NP formulations that can be used directly without any difficulty even after long term storage. Keeping this in view, recent research efforts have been focused to develop *in-situ* synthesis of metal nanoparticles within polymeric gel networks. Polymers/hydrogels or microgels can produce well-defined morphologies of nanoparticles inside the network structures, which are of great value in bio-medical applications because of their exceptional compatibility with biological molecules, cells, and tissues. Gel networks can enable to generate semiconductor, metal and magnetic nanoparticles inside the gel networks.

3.1. Polymer Chain Networks

Polymer-assisted synthesis of metal nanoparticles has received considerable attention because (a) the small concentrations of polymer or co-polymer are capable of stabilizing nanoparticles effectively through steric stabilization, (b) the polymer functional groups serve as both reducing and stabilizing/capping agent, (c) it is simple to vary the size of the nanoparticles by controlling the polymer/metal salt ratio, and (d) ease of preparation of metal-polymer nanocomposites. The progress of nanoparticles and nanostructural materials

has opened new opportunities for building up functional nanostructures.

Most of the functional polymers can act as templates for the synthesis of nanosized particles with reasonable stability. Both natural as well as synthetic polymers are extensively used for this purpose. For example, natural cellulose fibers could be used as nanoreactors and particle stabilizer, for *in situ* synthesis of metal nanoparticles from metal precursor solutions [41,42]. In addition, Se nanobelts have also been synthesized by using a cellulose reducing method at low temperature [43]. Cellulose acetate containing iron and copper nanoparticle composites has demonstrated catalytic activity in hydrogenation of olefins and co-oxidation reactions [44]. Ultrafine cellulose acetate-silver nanoparticles have been reported for antibacterial activity too [45]. Recently, Liu *et al.* [46] have successfully incorporated iron oxide nanoparticles in regenerated cellulose film, which displayed anisotopicity. The good alignment of nanostructures within the film appears due to shrinkage of the film while drying. Gold nanoparticles, when embedded in a natural polymer like dextran, have been used as glucose sensor. To immobilize dextran on gold nanoparticles the first step is to modify gold nanoparticles dispersion with 16-mercaptohexadecanoic acid (16-MHDA) in the presence of tween-20 to obtain –COOH groups on the nanoparticles [24]. Then, 2-(2-amino-ethoxy) ethanol (AEE) is allowed to react with -COOH groups using carbodiimide reaction where –OH functionality is achieved throughout on the surface of nanoparticles. These -OH groups are activated by epichlorohydrin and coupled with amino groups of dextran to produce dextran-coated gold nanoparticles.

Gum acacia or Gum Arabic (GA) is a natural polysaccharide extensively used as a stabilizer and reducing agent for the synthesis of nanoaprticles of 5-10 nm size [47-48]. Huang *et al.* [49] have reported that silver nanoparticles containing cellulose acetate film were prepared by dissolving silver nitrate and cellulose acetate in 2-methoxy ethanol and then boiling the resulting mixture in a fume hood.

Likewise, synthetic functional polymers or co-polymers are also effective in the preparation of silver nanocomposites. Poly (vinyl alcohol), (PVA) a versatile biocompatible polymer, is widely used to form metal nanoparticles. For example, PVA -silver nanoparticles composite film can be obtained by dissolving PVA solution with a suspension of silver nanoparticles via citrate reduction, followed by solvent evaporation [50]. Gao *et al.* [51] have synthesized silver nanoparticles by using a two-armed polymer consisting of a crown ether core i.e., [poly (styrene)]-dibenzo-18-crown-6-[poly (styrene)]. The complexation effect of crown ether embedded in the polymer with the Ag^+ leads to aggregations of polymer-Ag^+ followed by the Ag^+ ions reduction photo-chemically by visible light to form Ag nanoparticles. A similar methodology has also been employed for polyelectrolyte capsule-nanoparticles synthesis.

[Poly (styrene sulfonate) (PSS) (core)/poly(allylamine hydrochloride) (shell)] received much attention due to both practical and fundamental aspects. In detail, the controlled photo-chemical reaction of silver, under the same conditions occurs mostly inside the capsule of polyelectrolyte and at the same time a minute quantity of Ag particles could also be found on the shell portion, but it is not observed in the surrounding solution where there are no PSS polymeric chains present.

3.2 Hydrogel Networks

Three-dimensional hydrogel networks are relatively suitable for the fabrication of nanoparticles than the conventional polymeric template systems such as block copolymers, star copolymers, bio-macromolecules, dendrimers, liquid crystals, latex particles, and so on. The most important reward is that hydrogel networks offer free-s-pace between the cross-

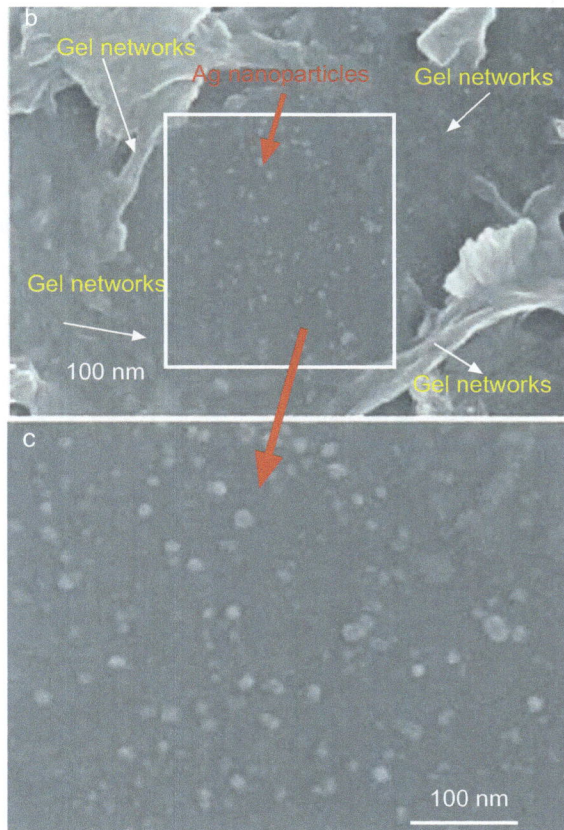

Fig. (1). SEM images of hydrogel networks as nano-reactors for nanoparticles. Images reprinted with permission from Ref. [52] Copyright © 2007 Elsevier.

linked chains networks in the swollen stage, which can serve for both nucleation and growth of nanoparticles. Fig. (**1**) clearly represents the nanoparticles formation throughout the cross-linked networks of hydrogel.

Mohan *et al.* [52] proposed a facile approach for obtaining ~3 nm sized silver nanoparticles within the poly(N-isopropylacrylamide-co-sodium acrylate) (PNIPAM-SA) hydrogel networks. In this novel methodology, PNIPAM based hydrogels are allowed to swell completely in distilled water, and then placed in $AgNO_3$ solution for Ag^+ ions to enter inside the swollen network and finally, transferred into solution of sodium borohydride ($NaBH_4$) for the reduction of silver ions to silver nanocomposite. These hydrogel-silver nanohybrids exhibited fair antibacterial activity, with the activity depending on the size of nanoparticles. In addition, this study clearly confirmed how the hydrogel networks could act as nano reactors for nanoparticles. An alternative approach [53], solution state polymerization of monomer acrylamide into polyacrylamide (PAM) hydrogel is carried out in aqueous medium containing Ag^+ ions. The Ag^+ ions functionalized PAM hydrogel matrix is then hydrolyzed to yield colloidal Ag nanoparticles within the hydrogel network. Thermosensitive PNIPAM has also been exploited to produce hydrogel coated gold nanoparticles [54] by surfactant free emulsion polymerization (SFEP) method. The formation of silver nanoparticles within the hydrogel networks occurs by ion exchange process. It is also possible to synthesize various forms of particles using hydrogel networks. Here, the hydrogel networks control the structures of the growing nanoparticles. A schematic illustration is presented in Fig. (**2**).

3.3 Microgel Networks

Microgels are generally defined as nanoscopic hydrogel network objects, which are essentially cross-linked macromolecule chains with a globular shape of size 50-100 nm. The functional microgels can be tailored to bear reactive functional groups, which are able to interact with metal ions. In this way, microgels can be loaded with silver precursors, which are then reduced within the microgel networks to yield silver nanoparticles. The surface of poly(styrene sulfonate)-doped polyaniline/poly(allylamine hydrochloride) capsules has also been used for photoreduction of silver ions to yield well-defined microgel particles containing silver nanoparticles [55]. The polystyrene core, grafted with thermosensitive poly(N-isopropylacrylamide) shell has also been exploited for the inclusion of Ag nanoparticles [56]. The inclusion nanosystem has led to the formation of a smart catalyst system. The catalytic activity of hybrid Ag nanoparticles for the reduction of 4-nitrophenol to 4-aminophenol by $NaBH_4$ can be tuned by temperature. Figure 2 illustrates the TEM images of the silver

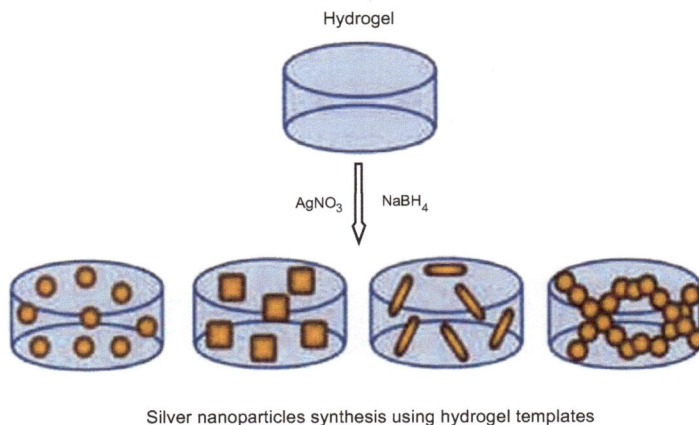

Fig. (2). Silver nanoparticles synthesis using hydrogel networks as templates.

nanoparticles throughout the functionality on the microgels poly (styrene)-poly (ethylene glycol methacrylate). Poly[vinylcaprolactum-*co*-(aceto acetoxtethyl methacrylate)] (VCL/AAEM) microgels, functionalized with Ag nanoparticles have been studied for their physio-chemical properties [57]. Incorporation of silver nanoparticles content results in the decrease of size of microgel particles. Moreover, the thermo responsiveness also decreases with AgNPs content. These hybrid microgel systems are excellent catalytic systems where AgNPs are separated from each other within the microgel network, thus providing an extremely large surface area. Localization of AgNP in microgel template allows easy separation from the reaction mixture. The functional groups, present in the microgel structure play a significant role in controlling the growth of nanoparticles. In a study, hybrid fluorescent microgels were prepared via photoactivated synthesis of Ag nanoparticles in PNIPAM microgels [58] Fig. (3). It was found that the presence of carboxylic groups in the microgel structure allowed the effective uptake of Ag$^+$ ions and controlled nucleation and growth of small nanoclusters.

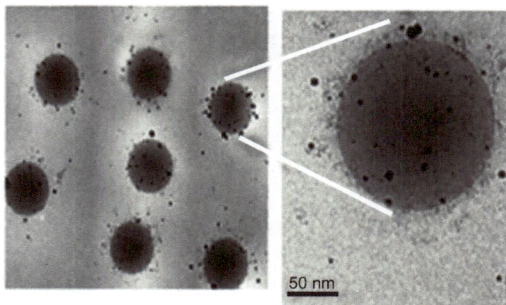

Fig. (3). TEM images for PS-PEGMA-Ag composite particles. Images reprinted with permission from Ref. [58] Copyright © 1999 American Chemical Society.

4. CONCLUSIONS

The incorporation of metal nanoparticles into polymer, hydrogel, microgel, nanogel network structures has

resulted a new generation of composite materials that have proved attractive interest in biomedical, catalytic, optical and electronic as well as quantum-size domain applications. This book chapter describes different approaches that have been implemented for the design and synthesis of metal nanocomposites with polymers and gel network structures.

5. REFERENCES

[1]   Sreekumaran Nair A, Pradeep T. Halocarbon mineralization by metal nanoparticles. Current Science 2003; 84(12): 1560-1564.

[2]   Cheng C , Chen Y, Lu C. Organic vapour sensing using localized surface plasmon resonance spectrum of metallic nanoparticles self assemble monolayer. Talanta 2007; 73(2): 358-365.

[3]   Wu Y, Li Y, Ong B, Liu P, Gardner S, Chiang B. High-Performance Organic Thin-Film Transistors with Solution-Printed Gold Contacts. Adv. Mater. 2005; 17(2): 184-187.

[4]   Huang D, Liao F, Molesa S, Redinger D, Subramanian V. Plastic- Compatible Low Resistance Printable Gold Nanoparticle Conductors for Flexible Electronics. J. Electrochem. Soc. 2003; 150 (7): 412-417.

[5]   Fuller SB, Wilheim EJ, Jacobson JM. Ink-jet printed nanoparticle microelectromechanical system. J. Microelectromechanical Syst. 2002; 11(1): 54-60.

[6]   Magdassi S, Bassa A, Vinetsky Y, Kamyshny A. Silver Nanoparticles as Pigments for Water-Based Ink-Jet Inks. Chem. Mater. 2005; 15(11): 2208-2217.

[7]   Jain TK, Morales MA, Sahoo SK, Leslie-Pelecky DL, Labhasetwar V. Iron Oxide Nanoparticles for Sustained Delivery of Anticancer Agents. Molecular Pharmaceutics 2005; 2 (3): 194 -205.

[8]   Khanna VK. Nanoparticle-based sensors. Defence Science Journal 2008; 58(5): 608-616.

[9]   Zhu Y, Sun D-X, Preparation of silicon dioxide/ polyurethane nanocomposites by a sol-gel process. J. Appl. Polym. Sci. 2004; 92(3): 2013-2016.

[10]   Sohn BH, Cohen RE. Electrical properties of block copolymers containing silver nanoclusters within oriented lamellar microdomains. J. Appl. Polym. Sci. 1997; 65(4): 723-729.

[11]   Bonnemann H, Richards RM. Nanoscopic Metal Particles - Synthetic Methods and Potential Applications. European Journal of Inorganic Chemistry 2001; 2001(10): 2455-2480.

[12]   (a) Creighton JA,. Blatchford CG, Albrecht MG. Plasma resonance enhancement of Raman scattering by pyridine adsorbed on silver or gold sol particles of size comparable to the excitation wavelength. J. Chem. Soc., Faraday

Trans.2, 1979; 75: 790-798 (b) Lee PC, Meisel D. Adsorption and Surface-Enhanced Raman of Dyes on Silver and Gold Sols. J. Phys. Chem. 1982; 86 (17): 3391-3395.

[13] Nickel U, Mansyreff K, Schneider S. Production of monodisperse silver colloids by reduction with hydrazine: the effect of chloride and aggregation on SERS signal intensity. J. Raman Spectros. 2004; 35 (2): 101-110.

[14] Li YS, Cheng JC, Coons LB. A silver solution for surface-enhanced Raman scattering. Spectrochemica Acta Part a-Molecular and Biomol. Spectr. 1999; 55(6): 1197-1207.

[15] Leopold N, Lendl B. A New Method for Fast Preparation of Highly Surface-Enhanced Raman Scattering (SERS) Active Silver Colloids at Room Temperature by Reduction of Silver Nitrate with Hydroxylamine Hydrochloride. J. Phys. Chem. B 2003; 107(24): 5723-5727.

[16] Chou KS, Lai YS. Effect of polyvinyl pyrrolidone molecular weights on the formation of nanosized silver colloids. Mater. Chem. Phys. 2004; 83(1): 82-88.

[17] Sondi I, Goia DV, Matijevic E. Preparation of highly concentrated stable dispersions of uniform silver nanoparticles. J. Colloid Interface Sci. 2003; 260(1): 75-81.

[18] Bright RM, Musick MD, Natan MJ. Preparation and Characterization of Ag Colloid Monolayers. Langmuir. 1998; 14(20): 5695-5701.

[19] Tan Y, Dai X, Li Y, Zhu D. Preparation of gold, platinum, palladium and silver nanoparticles by the reduction of their salts with a weak reductant–potassium bitartrate. J. Mater. Chem. 2003; 13: 1069-1075.

[20] Wang X, Zhuang J, Peng Q, Li Y. A general strategy for nanocrystal synthesis. Nature 2005; 437: 121-124.

[21] Sun Y, Xia Y. Shape-Controlled Synthesis of Gold and Silver Nanoparticles. Science 2002; 298(5601): 2176-2179.

[22] Green M, Allsop N, Wakefield G, Dobson PJ, Hutchison JL. Trialkylphosphine oxide/amine stabilised silver nanocrystals—the importance of steric factors and Lewis basicity in capping agents. J. Mater. Chem. 2002; 12: 2671-2674 .

[23] Kim SW, Paark J, Jang Y, Chung Y, Hwang S, Hyeon T. Synthesis of Monodisperse Palladium Nanoparticles. Nano Lett. 2003; 3(9):1289-1291.

[24] Bunge SD, Boyle TJ, Headley TJ. Synthesis of Coinage-Metal Nanoparticles from Mesityl Precursors. Nano Lett. 2003; 3(7): 901-905.

[25] Rodríguez-Gattorno G, Diaz D, Rendon L, Hernandez-Segura GO. Metallic Nanoparticles from Spontaneous Reduction of Silver(I) in DMSO. Interaction between Nitric Oxide and Silver Nanoparticles. J. Phys. Chem. B 2002; 106(10): 2482-2487.

[26] Rospendowski BN, Kelly K, Wolf CR, Smith WE. Surface-enhanced resonance Raman scattering from cytochromes P-450 adsorbed on citrate-reduced silver sols. J. Amer. Chem. Soc. 1991; 113(4): 1217-1225.

[27] Pillai ZS, Kamat PV. What Factors Control the Size and Shape of Silver Nanoparticles in the Citrate Ion Reduction Method? J. Phys. Chem. B 2004; 108(3): 945-951.

[28] Kumar M, Varshney L, Francis S. Radiolytic formation of Ag clusters in aqueous polyvinyl alcohol solution and hydrogel matrix. Radiat. Phys. Chem. 2005; 73(1): 21-27.

[29] Raveendran P, Fu J, Wallen SL. Completely "Green" Synthesis and Stabilization of Metal Nanoparticles. J. Am. Chem. Soc. 2003; 125(46): 13940-13941.

[30] Huang H, Yang X. Synthesis of polysaccharide-stabilized gold and silver nanoparticles: a green method. Carbohydrate Research 2004; 339(15): 2627-2631.

[31] Beveridge TY, Murray RGE. Sites of Metal Deposition in the Cell Wall of Bacillus subtilis. Journal of Bacteriology 1980; 141: 876-887.

[32] Campbell CT. Cs promoted Ag(111): model studies of selected ethylene catalysis. Journal of Phys. Chem. 1985; 89(26): 5789-5795.

[33] Darnall DW, Greene B, Henzl MJ, Hosea M, McPherson RA, Sneddon JJ, Alexander MD. Selective recovery of gold and other metal ions from an algal biomass. Environment and Science and Technology 1986; 20 (2): 206-208.

[34] Teng F, Xu JG, Tian ZJ, Wang JW, Xu YP, Xu ZS, Xiong GX, Lin L. Formation of a novel type of reverse microemulsion system and its application in synthesis of the nanostructured $La_{0.95}Ba_{0.05}MnAl_{11}O_{19}$ catalyst. Chem. Commun. 2004; 16: 1858-1859.

[35] Zhang W, Qiao X, Chen J. Synthesis and characterization of silver nanoparticles in AOT microemulsion system. Chem. Phy. 2006; 330 (3): 495-500.

[36] Kitchens CL, McLeod MC, Roberts CB. Solvent Effects on the Growth and Steric Stabilization of Copper Metallic Nanoparticles in AOT Reverse Micelle Systems. J. Phys. Chem. B 2003; 107(41): 11331-11338.

[37] Hota G, Jain S, Khilar KC. Synthesis of CdS–Ag_2S core-shell/composite nanoparticles using AOT/n-heptane/water microemulsions. Colloid. Surf. A 2004; 232(2-3): 119-127.

[38] Petit C, Lixon P, Pileni MP. In situ synthesis of silver nanocluster in AOT reverse micelles. J. Phys. Chem. 1993; 97(49): 12974-12983.

[39] Zhang WZ, Qiao XL, Chen JG, Wang HS. Preparation of silver nanoparticles in water-in-oil AOT reverse micelles. J. Colloid Interf. Sci. 2006; 302(1): 370-373.

[40] May A., Shaul B. Molecular Theory of the Sphere-to- Rod Transition and the Second CMC in Aqueous Micellar Solutions. J. Phys. Chem. B. 2001; 105 (3): 630-640.

[41] Yu F, Liu Y, Zhuo R. A novel method for the preparation of core-shell nanoparticles and hollow polymer nanospheres. J. Appl. Polym. Sci. 2004; 91(4): 2594–2600.

[42] Mucalo MR, Bullen CR, Manley-Harris M, McIntire TM. Arabinogalactan from the Western larch tree: A new, purified and highly water-soluble polysaccharide-based protecting agent for maintaining precious metal nanoparticles in colloidal suspension. J. Mater. Sci. 2002; 37(3): 493–504.

[43] Lu Q, Gao F, Komarneni S. Cellulose-Directed Growth of Selenium Nanobelts in Solution. Chem. Mater. 2006; 18(1): 159–163.

[44] Shim II-W, Choi S, Noh WT, Kwon J, Chao JY, Chae DY, Kim KS. Preparation of Iron Nanoparticles in Cellulose Acetate Polymer and their Reaction Chemistry in the Polymer. Bull. Korean Chem. Soc. 2001; 22(7): 772–774.

[45] Shim II-W, Noh WT, Kwon J, Chao JY, Kim KS, Kang DH. Preparation of Copper Nanoparticles in Cellulose Acetate Polymer and the Reaction Chemistry of Copper Complexes in the Polymer. Bull. Korean Chem. Soc. 2002; 23(4): 563–566.

[46] Liu S, Zhou J, Zhang L, Guan J, Wang. Synthesis and Alignment of Iron Oxide Nanoparticles in a Regenerated Cellulose Film. J. Macromol. Rapid Commun.2006; 27(24): 2084–2089.

[47] LaConte L, Nitin N, Bao G. Magnetic nanoparticle probes. Mater. Today 2005; 8(5): 32-38.

[48] Kwon JW, Yoon SH, Lee SS, Seo KW, Shim W. Preparation of Silver Nanoparticles in Cellulose Acetate Polymer and the Reaction Chemistry of Silver Complexes in the Polymer. Bull. Korean Chem. Soc. 2005; 26(5): 837-840.

[49] Huang J, Matsunaga N, Shimanoe K, Yamazoe N, Kunitake T. Nanotubular SnO_2 Templated by Cellulose Fibers: Synthesis and Gas Sensing. Chem. Mater. 2005; 17(13): 3513-3518.

[50] Khanna PK, Singh N, Kulkarni D, Deshmukh S, Charan S, Adhyapak PV. Water based simple synthesis of re-dispersible silver nano-particles. Materials Letters 2007; 61(16): 3366-3370.

[51] Gao J, Fu J, Lin C, Lin J, Han Y, Yu X, Pan C. Formation and Photoluminescence of Silver Nanoparticles Stabilized by a Two-Armed Polymer with a Crown Ether Core. Langmuir 2004; 20(22): 9775-9779.

[52] Mohan YM, Lee K, Premkumar T,Geckeler KE. Hydrogel networks as nanoreactors: A novel approach to silver nanoparticles for antibacterial applications. Polymer 2006; 48(1): 158-164.

[53] Saravanan P, Raju MP, Alam S. A study on synthesis and properties of Ag nanoparticles immobilized polyacrylamide hydrogel composites. Mater. Chem. Phys. 2007; 103 (2-3): 278-282.

[54] Kim JH., Lee TR. Thermo- and pH-Responsive Hydrogel-Coated Gold Nanoparticles. Chem. Mater. 2004; 16(19): 3647-3651.

[55] Antipov AA, .Sukhorukov GB, Fedutik YA, Hartmann J, Giersig M, Mohwald H. Fabrication of a Novel Type of Metallized Colloids and Hollow Capsules. Langmuir 2002; 18(17): 6687-6693.

[56] Lu Y, Mei Y, Drechsler M, Ballauff M. Thermosensitive Core–Shell Particles as Carriers for Ag Nanoparticles: Modulating the Catalytic Activity by a Phase Transition in Networks. Angew. Chem. Int. Ed. 2006; 45(5): 813-816 .

[57] Pich A, Karak A, Lu Y, Ghosh AK, Adler HJP. Preparation of Hybrid Microgels Functionalized by Silver Nanoparticles.Macromol. Rapid Commun. 2006; 27(5): 344-350.

[58] Cairns DB, Armes SP, Bremer LGB. Synthesis and characterization of submicrometer-sized polypyrrole-polystyrene composite particles. Langmuir 1999; 15(23): 8052-8058.

INDEX